ESSAYS ON CONSCIOUSNESS:

Towards a New Paradigm

INGRID FREDRIKSSON

BALBOA.
PRESS

A DIVISION OF HAY HOUSE

This book is a work of non-fiction. Unless otherwise noted, the author and the publisher make no explicit guarantees as to the accuracy of the information contained in this book and in some cases, names of people and places have been altered to protect their privacy.

Scripture taken from the King James Version of the Bible.

Balboa Press books may be ordered through booksellers or by contacting:

Balboa Press
A Division of Hay House
1663 Liberty Drive
Bloomington, IN 47403
www.balboapress.com
1 (877) 407-4847

Because of the dynamic nature of the Internet, any web addresses or links contained in this book may have changed since publication and may no longer be valid. The views expressed in this work are solely those of the author and do not necessarily reflect the views of the publisher, and the publisher hereby disclaims any responsibility for them.

The author of this book does not dispense medical advice or prescribe the use of any technique as a form of treatment for physical, emotional, or medical problems without the advice of a physician, either directly or indirectly. The intent of the author is only to offer information of a general nature to help you in your quest for emotional and spiritual well-being. In the event you use any of the information in this book for yourself, which is your constitutional right, the author and the publisher assume no responsibility for your actions.

Any people depicted in stock imagery provided by Getty Images are models, and such images are being used for illustrative purposes only.
Certain stock imagery © Getty Images.

Print information available on the last page.

ISBN: 978-1-9822-0811-0 (sc)
ISBN: 978-1-9822-0813-4 (hc)
ISBN: 978-1-9822-0812-7 (e)

Library of Congress Control Number: 2018908150

Balboa Press rev. date: 10/24/2018

ACKNOWLEDGMENTS

TO TONIETTA WALTERS

Adjunct Professor of Philosophy on Miami Dade College - Virtual College, Adjunct Professor of Philosophy at Broward College and Founder & Executive Director at Noumen Art Center for Applied Aesthetics. Previously: High IQWorld and College for Ultra Lyrical Abstraction.

Tonietta was supposed to co-author this book with me and add her own chapter of Spirit. We stayed in touch as she was writing and talked just a few days before her death and she was full of anticipation and she was so excited about the book. She hadn't felt very well recently though and a just a couple of days later the tragic message reaches me that Tonietta's heart didn't have the strength anymore and she had gently and calmly fallen asleep. Her mother Marjorie was with her at her hospital.

The world and my life will not be the same without Tonietta Walters, and neither will this book, but I am convinced that she and her spirit is with us all. She holds each and every one of us in her hands, she protects us and blesses this book. This is for you, Tonietta.

I am also very grateful to every contributors in this book for all their work with new ideas and Annica Andersson for the art on the cover.

To all, both named and unnamed, my heartfelt thanks.

CONTENTS

CONTENTS

INTRODUCTION

Y OU HAVE IN YOUR hand "Essays on Consciousness: Towards a New Paradigm," my third book on Consciousness. As time progresses, and my authorship too, I feel everything gets more and more exciting. The first essay in the first book "Aspects of Consciousness: Essays on Physics, Death and the Mind" was about Remote Viewing, written by Russell Targ and Elizabeth A. Rausher. In the second book "The Mysteries of Consciousness: Essays on Spacetime, Evolution and Well-Being" was Menas Kafatos one of the authors with "The Non-local Universe is the Conscious Universe", which Deepak Chopra co-authored. In this even more thrilling book, there are, as usual, well-known and famous co-authors from whole the word, e.g. **Rupert Sheldrake**. He asks "Where is the mind?" and "If there were no people and no animals, would there be minds?"

Animals also have telepathic abilities; dogs and cats know when the owner is coming home. In Los Angeles and Santa Cruz Cruz fifty percent of all dogs, if not more, know when their owners are coming home. Rupert Sheldrake also talks about the Morphic fields, the organizing fields.

Olle Johansson writes about some more dangerous effects. Do not believe that mobile phones, wireless internet and so on are safe! They interfere with normal brain function, learning and memory, fertility and cancer risks and have been shown to shatter the DNA in cells. All of this can be found in peer review-based scientific journals, but up to now have not been in the public domain and very few will tell you about it. Very few will try to protect you and very few want to speak the truth.

Attila Grandpierre writes about the Absolute, the ultimate fix point of the world, the self-existing entity on which we can stand our footing firmly. To reach that aim we consider the relevant achivements of the last millenia in the light of the enlarged, postmodern science that preserves all the power of physics yet generalizes it to take into account the own laws of life and consciousness. In our search we are led to recognize the essentially complete view of the Universe consisting from three levels of reality: phenomena, laws and first principles. We have found three first principles of the Universe, that of matter, life and consciousness. We show that the architecture of the Universe has a fundamental triune structure corresponding to the first principles of physics, biology and consciousness. We point out that this fundamental Trinity of the Universe is the product of a still deeper, unmanifested entity. We found the ultimate, self-existing reality what we call as the One, the unified whole of the life principle, the Cosmic Self and the consciousness principle.

G Anita K Westlund has found the holy numbers on our Earth. She writes about the numbers ever repeated are 144, 233 and 377, the same as manifested in the ancient Cheops pyramid. This is hardly not believable if not viewing the missing fact of chakra effecting us, involved in the Fibonacci number of 377. How is it going and why, asks G Anita K Westlund. She also talks about sounds.

We need music to feel well. **Alexander J. Graur** writes about Music and Consciousness. Like Music, Consciousness is a complex phenomenon. Let's consider a perfectly cut diamond. How could we define it? Is it a stone? Certainly. Is it made by carbon? Mostly. Does it have any practical value? The evaluation depends on where you look, what you are looking for, what are you looking to, under what type of light (artificial or natural), why are you looking for, what you intend to do with it, and so on. I do not believe we will ever have a comprehensive definition of the consciousness. But the quest itself is a proof that consciousness exists.

Klaus Stüben writes about The open door to realize limitless consciousness and The dilemma of science investigating consciousness. "What if consciousness is not something inside us, like a force that we have and need to wake up and use, rather something that we are a part of. What if there is nothing else but consciousness, no matter what is showing up at any moment. What if we are also only showing up in that consciousness as a product of consciousness – showing up in whatever form, whatever feeling, whatever emotion and sensation. What if consciousness can only be experienced in a direct way. As it will be shown in case examples in this article as well as experienced in different experiments the qualities of consciousness can also be experienced directly".

The Origin and History of the Human Mind by **Carl Johan Calleman** is about the nature of consciousness and the relationship between consciousness and matter. It have been discussed for a very long time. In this debate, there has been two conflicting schools of thought; the materialist and the idealist. The materialists have held the viewpoint that everything that exists is fundamentally matter and that the workings of the human mind can be understood from chemico-biological processes in the brain. On the other side of the fence have been the idealists who at its extreme end have held the viewpoint that physical reality is merely an illusion and that all matter is an expression of consciousness. In addition, there is the so-called dualist viewpoint, which sees matter and consciousness as two separate and yet interacting aspects of the universe. Dualism was notably elaborated by Descartes at the beginning of the scientific revolution in the early 1600's and is most likely the predominant viewpoint today.

Eve Isham is a young experimental psychologist. She says: I must emphasize the two adjectives "young" and "experimental." Young – meaning I am only starting out in my career and I am still sorting things out in my head. But this also means that I am open to new ideas instead of being married to a theory that desperately begs for a re-evaluation. "Experimental" reflects my scientific training and my appreciation for the value of careful experimental designs. I value conservative conclusions and scientific approaches that require the exhausting of all possibilities before claiming causation over correlation. Having said that, here is my story about how my graduate advisor, Bill Banks, and I tried to save free will.

Gerard J.F. Blommestijn have been on "every" conference about consciousness. He writes about the quantum mechanical description of nature. In the quantum mechanical description of nature one distinguishes two quite different processes: 1) the completely deterministic evolution of the wave function in the absence of measurements, and 2) the measurement process itself, during which the wave function reduces in a fundamentally unpredictable way to one of its components corresponding to one of the possible values of the measured observable. There are reasons to suspect that this quantum mechanical indeterminacy plays a role in the brain processes that are related to consciousness. In this article a theory is presented on the function of quantum mechanical wave function reduction in relation to consciousness.

Ingrid Fredriksson asks if consciousness influence our epigenetics and can epigenetics influence our consciousness? Epigenetics is a mechanism for regulating gene activity independent of DNA sequence that determines which genes are turned on or off: in a particular cell type, in a different disease states or in response to a physiological or even psychological stimulus.

There is a microbiota–gut–brain axis communication in health and disease. The molecules that constitute epigenomes have no resemblance of DNA. While DNA is a double spiral, similar to a twisted rope ladder, the epigenome is a system of chemical markers that sits on the DNA. What is its purpose? In the same manner that a conductor leads an orchestra, the epigenome decides how the genetic information of DNA shall be should be expressed. The molecule markers either engage or disengage the genes depending upon the cell's needs and environmental factors, such as diet, stress and poisons. Of late, the discoveries surrounding the epigenome have caused a revolution in the field of biology now being able to prove a connection between the epigenome and certain illnesses, including aging.

Richard L. Amoroso writes about Millennial Science, The Imminent Age of Discovery's 'Conscious' Technologies. A new 'Age of Discovery' inclusive of the 'Physics of the Observer' (nature of awareness) – discovery of the mind or consciousness is pragmatically imminent. A comprehensive noetic theory only awaits completion of empirical tests that like the discovery of electricity leads to a plethora of technologies – 'Conscious' technologies. A 'noetic effect' transducing physically real Cartesian mind-body interaction principles inherent in Einstein's Unified Field (U_F) is required. Quantum computing (QC) designs modeled to include U_F mind-body interaction parameters, operate the systems. Discovery of the mind is briefly reviewed, salient technology summarized; then favorite 'spirit-based' medical applications.

TO UNDERSTAND ADVERSE HEALTH EFFECTS OF ARTIFICIAL ELECTROMAGNETIC FIELDS… …IS "ROCKET SCIENCE" NEEDED OR JUST COMMON SENSE?

Olle Johansson, PhD, associate professor
The Experimental Dermatology Unit, Department of
Neuroscience, Karolinska Institute, 17177 Stockholm,
Sweden (due to retirement, new address since November
15, 2017: Rådjursvägen 3, 130 56 Utö, Sweden)

AS WE ARE ALL RAPIDLY BEING FORCED into the new generation of electronic gadgets and wireless services, sometimes referred to as the "Internet of Things" and primarily based on the 5th generation (5G) of wireless communication, to be followed soon by 6G and 7G, more and more people are asking themselves if the ever-increasing levels of artificial electromagnetic fields (EMFs), especially of the pulsed type, really are safe for living organisms; if their various equipments are there only for them to enjoy as private persons or for political, military, and commercial surveillance purposes – it is already obvious that it may easily develop into the "Internet of Totalitarian Control", with unbelievably dark aspects of artificial intelligence, human brain control, and the Digital New World Order just around the corner; if some gadgets add fire risks; if they violate integrity considerations; if they pose an open door for advanced cyberattacks of homes, schools, and workplaces; if they generate profit for private companies but mere costs for the public health and welfare system; and if their EMFs pose a threat to their own and their families' health, as well as to the environment. Already at the end of the 1970s, I decided that this is an issue that needs to be looked at much more closely and perhaps even needs—for many serious reasons—to be stopped before it turns on us and other animals, plants, and bacteria as the *Idiocy of Things!*

Over the years, more and more people all over the world have become concerned citizens who realize that modern electronic devices produce electromagnetic fields that are not native to our Planet. These non-native EMFs are completely foreign to human, animal, and plant biology and may be wreaking havoc on the health and well-being of humanity and other life forms on our Planet.

In contrast to those with these fears, various opponents continue to insist that the majority of studies show no risk. But so does the majority of all car journeys, and still people die in car accidents! I am very happy that car manufacturers turn their backs on the nearly 100 percent of car journeys that end happily and instead concentrate on the trips that result in injury, disability, and death, with the intention to make their cars safer and safer!

I have wondered my whole professional life why certain persons—especially those with political and economic power—seem to be completely blind, while anyone in the street sees the problem and the solution right away. Investigative journalism has, of course, revealed that now and then, people who can't see the obvious have been blinded by profit and greed, power and money or led astray by ice-cold marketing, including RFID (radiofrequency identification) chipping of Belgians and Swedes (see http://www.dailymail.co.uk/sciencetech/article-4203148/Company-offers-RFID-microchip-implants-replace-ID-cards.html and http://www.abc.net.au/news/2017-04-03/swedish-employees-agree-to-microchip-implants/8410018).

Against this background, I decided a few years ago to inaugurate *"The Institute of Common Sense for Common Sense"* since I do not believe the question of possible health and environmental effects of all these artificial electromagnetic fields boils down to any type of advanced, "rocket science"-type thinking but is just a matter of using our brilliant brains to ponder the medical and biological impacts and how to easily and rapidly solve these problems. The number of scientific papers giving us the key elements of this knowledge is increasing rapidly by the day and has already far surpassed what we actually need to act with precision in the name of the Precautionary Principle and sound, sensible risk management (Dämvik and Johansson 2010).

The simplicity of this general issue — which in parallel also has resulted in, and is still resulting in, hundreds of fruitless meetings as well as in many expensive and downright harmful articles in world-famous journals being sponsored by the vested interests of industry and finance — is that modern, artificial electromagnetic fields must be regarded as a *highly toxic environmental exposure*, something I have pointed out in public a countless number of times. I wish I could say that there soon will be an end to this full-scale, 24/7 human, animal, plant, and bacteria experiment, the largest ever on this Planet, which every organism definitely has not been informed of or given any consent to, completely in contraposition to the Nuremberg Code of Ethics of 1947. But I can not, not even with the alarming results of the recent American National Toxicology Program (NTP) cancer study that has shown that rats exposed to mobile telephony for two years have an increased incidence of aggressive brain tumours (gliomas) and malignant heart tumours (schwannomas) (Wyde *et al* 2016, manuscript: Report of Partial Findings from the National Toxicology Program Carcinogenesis Studies of Cell Phone Radiofrequency Radiation in Hsd: Sprague Dawley® SD rats (Whole Body Exposures). http://biorxiv.org/content/biorxiv/early/2016/05/26/055699.full.pdf). Oddly enough, the media in Scandinavia and in many other countries have barely covered this study. Is some form of ongoing, Planet-wide white- or green-washing happening?

3

This NTP project has been underway for more than a decade and, with a $25 million price tag, is the most expensive ever undertaken by the NTP. It involved more than 2,500 rodents exposed for nine hours every day for two years to the same type of radiation and frequencies found in cell phones.

In addition to these cancer-incidence data, the American National Toxicology Program has recently revealed that the same radiofrequency/microwave radiation that led male rats to develop brain tumours also caused DNA breaks in their brains (http://microwavenews.com/news-center/ntp-comet-assay). These findings are part of the same $25 million NTP project. The NTP results provide *"strong evidence for the genotoxicity of cell phone radiation,"* professor Ron Melnick, who initiated the study, told *Microwave News*. This *"should put to rest the old argument that RF radiation cannot cause DNA damage,"* he said (see http://microwavenews.com/news-center/ntp-comet-assay).

Instead of intensifying research efforts with the above alarming 'luggage in hand', my own university, the famous Karolinska Institute, has chosen to go in the opposite direction. (My personal reflection on this is that with the above NTP results at hand, the need to move forward is much, much greater than ever!) Hopefully, their decision to back off will be, in the future, regarded as a wise and correct move. But the risk is that the current, very costly Paolo Macchiarini scandal—in which a visiting Professor at Karolinska Institute was performing trachea surgery that resulted in six out of the eight patients dying, ending in a misconduct scandal—will be followed by yet another one, affecting many more people (just the above-mentioned American National Toxicology Program cancer study outcome indicates that cell phone radiation could result, in the decades to come, in an additional 75–150 million (!) extra human cancer cases worldwide, or even many more. And with future huge numbers of unnecessary premature deaths not easily swept under the carpet. Thus, for academia and society to turn their backs on these new findings would be as daring as turning your back to an Egyptian cobra.

The NTP scientists must have regarded their results as highly important since they released them before the entire study was completed, a rather unusual decision. Their results have the potential to move a debate that has been locked in stalemate for almost as long as cell phones have been around. To say that the American NTP study is a paradigm-shifting one is to understate its importance.

In addition, very recently European investigators at the Ramazzini Institute in Italy arrived at similar conclusions. In their study they investigated radiofrequency effects in nearly 2,500 rats from the fetal stage until death (Falcioni *et al* 2018). As pointed out by Charles Schmidt in a Scientific American article early in 2018 (cf. https://www.scientificamerican.com/article/new-studies-link-cell-phone-radiation-with-cancer), it is also noteworthy that the two different studies, from the USA and from Italy, respectively, evaluated radiation exposures in different ways. The NTP looked at near-field exposures, which approximate how people are dosed while using their own cell phones. The Ramazzini researchers, however, looked at the far-field exposures, which approximate the wireless microwave radiation that hits us from sources all around us, including wireless devices such as Wi-Fi routers in schools, homes, workplaces and public spaces, smart meters, baby alarms, tablets and laptop computers. Yet the two studies generated comparable results with male rats in both studies developing Schwannomas of the heart at statistically higher rates than control animals that were not exposed.

Taken together, the findings "confirm that RF radiation exposure has biological effects" in rats, some of them "relevant to carcinogenesis," says Jon Samet, a professor of preventive medicine and dean of the Colorado School of Public Health, who did not participate in either study, but where interviewed by Charles Schmidt in his Scientific American article (cf. https://www.scientificamerican.com/article/new-studies-link-cell-phone-radiation-with-cancer).

Since 2011 radiofrequency radiation has been classified as a Group 2B "possible" human carcinogen by the International Agency for Research on Cancer (IARC), an agency of the World Health

Organization. Based on the new animal findings, and previous epidemiological evidence linking heavy and prolonged cell phone use with brain gliomas in humans (cf. Hardell *et al* 2007), professor Fiorella Belpoggi, director of research at the Ramazzini Institute and the Italian study's lead author, says IARC should consider changing the radiofrequency radiation designation to a "probable" human carcinogen (Group 2A). Even if the hazard is low, billions of people are exposed, she says, alluding to the estimated number of wireless subscriptions worldwide. According to Charles Schmidt, at Scientific American, Véronique Terrasse, an IARC spokesperson, says a reevaluation may occur after the NTP delivers its final report.

Already in 2001, researchers in Australia had reported one of the first scientific hypotheses that normal mobile phone use can lead to cancer (French *et al.* 2001). The research group, led by radiation and cell biology expert Dr. Peter French, at that time principal scientific officer at the Centre for Immunology Research at St Vincent's Hospital in Sydney, said that mobile phone frequencies well below current safety levels could stress cells in a way that has been shown to increase susceptibility to cancer. The paper, published in the June 2001 issue of the science journal *Differentiation,* concluded that repeated exposure to mobile phone radiation acts as a repetitive stress leading to continuous manufacture of heat shock proteins within cells. Their coauthors included Professor Ron Penny, then the director of the Centre and one of Australia's leading experts in the cellular effects of HIV, and Professor David McKenzie, the head of applied physics at Sydney University, all having genuinely good scientific reputations.

As recently pointed out to me by Dr. Lauraine Vivian who is an Honorary Research Associate at The Research Unit for General Practice and Section of General Practice, Department of Public Health, Faculty of Health and Medical Sciences, University of Copenhagen, in Denmark, we may soon run the risk of having few or even no one left to collect statistical data or to give treatment for tumours because national health systems are collapsing due to increases in cancer rates and similar health and infrastructure problems. Shall we then have to trust future robot doctors to treat the numerous cancers with their controllers locked in EMF-free cages, or shall we start today to seriously discuss the questions in front of us? And since cancer is not at all the biggest and scariest result of this full-scale experiment, I say that we must sit down now!

It is to be hoped that this mind-boggling scenario is mere science fiction, but it is fascinating to compare my employer's recent statement that *"there is no need for your services, Olle,"* with the reality that I could employ many, many people just to deal with all the daily needs of, and questions from, people around the planet, and perform laboratory and field work, and many other important tasks. A pretty ironic difference, if you ask me...

It is now more important than ever that, hand-in-hand, we all jointly embark on a journey to change the current research and health-care paradigms so that everyone feels the utmost confidence in us and our altruistic aspirations. The clients of a governmental scientist or county council-employed medical professional can never be misunderstood—they are the citizens, no others. We should consider no wallets, no CVs, no political considerations or ambitions, no spin-off companies, no options, no private profits, no industrial economic gain—nothing other than public health. Personally, I – as also proclaimed by the eminent cancer specialist professor Lennart Hardell in Örebro, Sweden - very strongly believe in that scientists have ethical and moral obligations never to turn their backs to any given cobras, even if it means that commerce will loose ground. ...Or do you not agree ...?

My personal journey started in the late 1970es when the first cases of what eventually was going to be termed the functional impairment (formerly disability or handicap; cf. Johansson 2015) electrohypersensitivity appeared, first in Norway and the USA, and later in other countries around the world, today with Taiwan,

Germany and Switzerland at the top, and with countries like Sweden at the bottom of the incidence scale. Often nowadays I look upon the electrohypersensitive persons as being the normal ones, with normal and natural biological avoidance behaviours, and the rest of the population (whom we today call "normal") actually being electro*hypo*sensitive. By being it, one certainly risks to become *i.a.* a number in the cancer statistics or infertility column (cf. below). The addition of the NTP and Ramazzini studies outcome is - of course - of paramount importance for this reasoning.

When I recently gave the talk, at the 10th Biennial European Conference of the Society for Scientific Exploration - in collaboration with the Swedish Society for Psychical Research (SSPR) and the research center Agora for Biosystems at the Sigtuna Foundation – named "Life and Mind – Scientific Challenges", Sigtuna, Sweden, Oct 13 – 15, 2016, which forms the basis for this book chapter, a person in the audience expressed a strong and angry disappointment that my lecture had not only contained science - since "the conference was supposed to be about science" - but I had also touched upon political arguments and references. I found it to be a surprisingly naive comment, especially from a scientist, since *politics is inherently present in everything and has proven to be the disguised hand that rocks the world*. There are no pure scientific projects any longer, and I say it has never been. Science is soaked in political and economical power struggle considerations, and to blind yourself to that fact will only lead you astray.

It is high time that we all, scientists, politicians, civil servants and citizens, finally realize how potentially dangerous man-made, artificial electromagnetic fields released from, and used by, our various electric and electronic gadgets – such as powerlines, transformers and wiring inside household items, cell phones, DECT phones, tablets, laptops, game centers, information tools, Bluetooth accessories, baby alarms and monitors, and gas, water and electricity wireless smart meters, may be for our health. If the opposite should be claimed with certainty, then all of the relevant published reports –now counting more than 26,000 in number according to EMF-Portal (https://

www.emf-portal.org/en) – all must be wrong at the same time, and the probability for that is – to say the least – infinitely small!

Some organizations that definitely, and to 100%, trust the current scientific results and our common knowledge about potential health effects of man-made, artificial electromagnetic fields, are the manufacturers, the operators, the radiation protection authorities and even the World Health Organization since they all (cf. below) have abandoned ship years ago. It is also, of course, no surprise that electromagnetic radiation no longer are covered by insurance as a result of health problems. The British insurance giant Lloyd's – together with other insurance and reinsurance companies – has launched a very vigilant move. Damage to health due to direct or indirect exposure to the electromagnetic radiation of our modern gadget-driven world are no longer covered by their insurance policies. So do not call the insurance companies in the future if you have become ill or sick due to mobile phone radiation, or if your child has come down with childhood leukemia due to powerfrequent magnetic field exposure, or an aggressive brain tumour or malignant heart cancer due to cell phone or Wi-Fi tablet radiation, since your health insurance does not cover it. You better look for the telephone number to your government and parliament since they allowed the public blanketing roll-out of these exposures. So you will have to - in the future - sue your government and parliament, meaning you will sue yourself since these administrative structures of society use *YOUR* tax money to cover their backs. In addition, critical whistleblowing scientists, casting long and large shadows of doubt on these so-called "safe" gadgets, have effectively been removed, instead of supported, so the roulette is right now spinning. But will it end as a Wheel of Fortune or will it end in disaster?

It would be highly suitable to follow the legally-based demands on the pharmaceutical industry and add an information leaflet to each wireless gadget sold telling the buyer that it has unwanted side-effects (and for which they have *i.a.* applied for technical patents based on cancer risks), some rare and some more common, but all potentially serious to your health. In the voice of democracy, we should also – like the tobacco industry has been forced to do – label each package with a warning informing the user that it may harm their health, that their insurance does not cover such damages, and that the manufacturers themselves tell you to keep it at least one inch from your body. Very early, I even suggested that it should be required by manufacturers, operators and radiation protection authorities that they would take on a personal responsibility - legally watertight - when they said that radiation is harmless. So far, no one has volunteered to sign such a personal responsibility legal contract. (I wonder why not…?) Instead they have gone to bizarre length to legally protect themselves from future liability claims and law suits over their "safe" products! This does not make sense to me; does it to you? No, it rather smells like quite another form of "precautionary principle", aimed at protecting the 'major players', and not the consumers.

A Belgian-Swedish study by Cammaerts & Johansson (2013) on ants, that were made unable to leave their artificial laboratory home, revealed that when exposed to cell phone radiation, the adult ants displayed obvious behavioral disorders, with more disruption in their daily activities and an increasingly scanning of their local environment. It was clear that something concerned them. I immediately after our 2013 study wrote a commentary in 2014 (https://takebackyourpower. net/experts-and-doctors-warn-pregnant-women-and-children-wireless/) where I urged pregnant women and children not to expose themselves to wireless radiation, and concluded that *we humans are mostly just standing around talking about this, whereas ants and bees are fleeing the field!* In it I also pointed to that a survey carried out in 2011 in Lausanne, Switzerland, had shown that the signal from the cell phones may not only confuse bees, but also cause their death. When researchers exposed beehives to cell phone radiation, the bees occupying the hive simply choose to move away and never return. I concluded that this is exactly the behaviour that beekeepers worldwide call CCD, Colony Collapse Disorder, a phenomenon that involves an abrupt disappearance of bees from their hives. Many other studies have in addition shown that bees are affected by and react to radiofrequency radiation. Scientists opine that exposure disrupts the hive, interferes with navigation, weakens the immune system (also cf. Johansson 2009a) and contributes to colony collapse (for references and further discussion, see Cammaerts 2017), so my idea above did find good ground for further exploration.

Thus, there is a real risk that democracy, nature, our habitat, garden ants, honey bees, etc., will be destroyed just because we are not watching them well enough... Instead we are giving away all our sex life secrets via various social media and universal 'clouds', in parallel to the authorities snooping around on the Internet (as part of the whole "social media" / Internet surveillance / stalking / troll-spreading / astroturfing / harassment / slander / bullying-bot carousel), we are watching our iPhones, updating our Facebook profiles, adding likes, checking Instagrams and YouTube channels. But can we really afford this constant ignorance? Can we allow ourselves to walk the Planet via our new VR glasses, instead of paying real attention? (Personally, I use my own eyes with ordinary spectacles - the world is then always in 3D, and with surround sound, surround smell, and a number of surround touch sensations. My personal slogan is *"Rather surround than surreal!"*. But am I enough numerous for Nature to survive the current toxic, man-made, electromagnetic field exposure; do I not have to depend on you too, dear Reader?)

Along these book pages, it will become clear that:
I want to discuss <u>*something*</u> which, in the form of Wi-Fi-enabled tablets and mobile phones, the educational authorities claim will revolutionize teaching and learning in spite of the fact that they have been used for years in Sweden parallel to an enormous drop in pedagogic quality and learning capacity, a fact brought up several

times during the recent political party leader's debates in the Swedish public service TV and radio as well as in newspapers/tabloids.

I want to discuss *something* which in the form of smart meters and high-frequency, low-energy, light bulbs is said to cut the energy costs, while our electricity bills constantly increase. (Also always remember that smart meters – wired or wireless - would have been, of course, a wet dream for Gestapo and SS giving Anne Frank and her family on Prinsengracht 263 in Amsterdam definitely no place to hide during WWII. The moment they had put on a single light bulb, they immediately would have been spotted and caught.)

I want to discuss *something* which many Waldorf-Steiner schools and Waldorf-Steiner daycare centers have decided not to welcome in their premises, since they do not want to have the access to mobile phones interfering with the educational work. They want students to be 100% present – not just their bodies but also their full attention – at their activities in class, since presence is essential for their educational practices and ideas. They also respect the need to adapt the school environment for staff and pupils that are electrohypersensitive. In such a way everyone is saved from any long-term health effects, such as impact on the fertility or cancer progress. Therefore, they have also, for the same reason, chosen not to have wireless networks in classrooms.

I want to discuss *something* which the insurance companies around the world, including "Lloyds" in the UK and "Swiss Reinsurance Co. Ltd." in Switzerland, refuse to take responsibility for. Among such items you find not only health effects of electromagnetic fields, but also health effects of GMO and nanotechnology. (…Is this not strange, they are all sold to us as 100% risk-free and completely safe. Then they should be safe to insure…!?)

I want to discuss *something* which the telecom manufacturers and operators completely and totally refuse liability for. Their products are safe, so they claim, but they do not – legally – touch them even with a barge pole or a pair of pliers. So, in a sense, these companies have their own precautionary principle (cf. above and below).

I want to discuss _something_ which the telecom manufacturers – for health safety reasons – tell you to keep at least one inch from your body.

I want to discuss _something_ which the radiation protection authorities around the world say is completely and totally safe, but – for safety reasons – still suggest that we shall use as little as possible, and to use a hands free accessory. Again, odd. Either a gadget is safe … or it is not.

I want to discuss _something_ which the radiation protection authorities around the world say gives off "very weak electromagnetic fields," in spite of the fact that a single mobile phone to your head has magnetic fields equal to lifting several electric train engines to the very same head.

I want to discuss _something_ which the radiation protection authorities around the world say gives off "very weak radiation", in spite of the fact that the current allowed public microwave exposure levels – compared to the natural background – are one quintillion (1,000,000,000,000,000,000) times stronger.

I want to discuss _something_ which easily penetrates walls, floors, ceilings … and you! And while penetrating you it feeds colossal levels of energy into your body, making molecules break, changing the behaviour and molecular machinery of your cells, damaging cells all the way to cell death, and feeding cell growth.

I want to discuss _something_ which lacks any form of biologically-based exposure standards or hygienic safety levels. Instead the safety of these gadgets are determined using so-called technical recommendations based on acute heating of fluid-filled plastic dolls, and only allowing you to make one (!) single, 6-minute long, mobile phone call once in your life-time. Serious? No! (Does that make you feel safe?)

In our modern world, we and our children are constantly flooded by various wireless devices, wherever we live, work, go to school or play. Many questions have arisen regarding whether the radiation is without harm, especially since the 'major players' clearly points out – through their complete refusal of liability - that wireless technology is NOT without risk. (So, as usual, the corporate industry conducts the classical tune: "Follow the money". The only question for them is always who is going to pay for the damages in the future and, for health effects of electromagnetic fields, they will not do it. More telling than any test tube or laboratory rat experiment ever may be, I would say...)

Numerous studies and reports, expert statements and overviews correctly states that "there is a strong suspicion of harm", and calls for the use of the Precautionary Principle as originally given by the 1992 Rio Declaration on Environment and Development. The World Health Organization (WHO) has thereafter classified powerfrequent magnetic fields (by 2001) as well as radiofrequency electromagnetic fields (by 2011) as possibly carcinogenic (2B), and recently, in 2012, the Italian Supreme Court ruled that mobiles can cause a brain tumour. Thus, we can immediately cross out the idea that these techniques would be safe, not even the WHO believes it – and they still have a category into which such proven safe exposures would fall ("Class 4 – proven human non-carcinogen"). The question now is instead how big the risk is and what we accept the risk to cost. Instead of avoiding the issue, it's high time to be completely outspoken, blunt, even to the point of rudeness, and to call things by their proper names without any 'beating about the bush'. To guarantee our and the Planet's health this is the only way forward. *I say loudly: call a spade a spade, please!*

Recently, in addition, a bill to ban phones in schools was introduced in **France** in 2009, and further tightened in January 29, 2015. Bans came into effect in places like **Nigeria** in 2012, around the time that teachers in the **Solomon Islands** called for phones to be banned in their schools. **Uganda** banned phones in schools in 2013, one year after **Malaysia** reaffirmed its own similar ban. And it's not only been in schools where young people have been prohibited from using their phones over the years. In one prefecture in **Japan** in 2014 children were not allowed to use phones after 9 pm, not long after the government in **Belgium** has announced measures to restrict the use of mobile phones by young children, sales of mobile phones to children under 7 years will be banned in shops and also on the internet, and adverts for mobile phones during children's programmes on TV radio and the internet will also be banned. In 2015, bans on student use of phones inside and outside of schools were considered in **Indonesia**, and in 2013 in **South Korea** experts have noticed a surge in teenagers with poor memory. This new ‹dementia› causes deterioration in cognitive abilities more commonly seen in people who have suffered a head injury or psychiatric illness. Experts blame game consoles and mobile phones for this worrying trend. Furthermore, a press release by February 27, 2017, has just been sent out about **Maryland** State's Children's Environmental Health And Protection Advisory Council being the first in USA to issue new recommendations to reduce Wi-Fi exposure of children, and a similar press release from March 6, 2017 tells us that "**Cyprus** Removes Wi-Fi from Kindergartens and Halts Wireless Deployment Into Public Elementary Schools". As a Swede I must, however, strongly wonder: in all these impressive statements and decisions … where is my own country Sweden? We cover our children in strongly coloured overalls, reflective vests, safety helmets, and more, but allow them to walk 'naked' in relation to the ambient artificial electromagnetic fields which have been cancer-classified by the WHO for nearly two decades!

All of the above should also be viewed against the very important notion that EMFs and autism in children may very well be associated. The first scientist to point to this was dr. George Carlo and his coworker dr. Tamara J. Mariea who in their 2007 article (Mariea & Carlo 2007) concluded that "*These data also suggest that wireless device EMR is a synergen in the etiology of autism, acting in conjunction with environmental and genetic factors, and offer a mechanistic explanation for the correlation between concurrent increases in the incidence of autism and the use of wireless technology*". Their ideas where very elegantly picked up by professor Martha Herbert and dr. Cindy Sage in their two, already classic, parallel papers in 2013 (Herbert & Sage 2013a,b), and have also built the foundation for the excellent public appearances of the entrepreneur and philanthropist Mr. Peter Sullivan (see e.g. http://

www.motherjones.com/environment/2016/12/silicon-valley-cellphones-wifi-sickness-emf-hypersensitivity), who has decided to use his economic means for the welfare of future generations.

All these gadgets are – from an evolutionary point-of-view – toys. Children who do not get tablets and smart phones still will mature to responsible and loving citizens - that you do not need to worry about! – but without the real life necessities such as clean water, clean air, food that can be eaten without risk, care, concern, love and respect, they will perish. As they also will if their sperm cell number is lowered beyond repair (cf. below).

As I pointed out in my recent article (Johansson 2016), 3.5-3.8 billion years ago the first unicellular organisms were formed and life was initiated on our Planet. During the coming thousands of million years cells divided to form multicellular plants and animals, and they grew more and more complicated and sophisticated. Soon our Planet was inhabited by insects, reptiles, fish, birds and finally mammals. During some 5,000,000 years mankind has made it's ascent, and our present subspecies, the *Homo sapiens*, has been around for about 200,000 years.

As stressed in the very same article, the recent massive roll-out of various wireless technologies should be critically viewed against this background. The last 100 years we have very suddenly been exposed to radio, TV, computers, cellular telephones, wireless internet, light ray tubes, compact fluorescent lamps, and house-hold appliances of various kind. And, as pointed out many times during the last four decades by myself, this is the actual central question: *"Can we really count on Darwinian evolution to ensure that our cells have developed an automatic protective shield against power-frequent electric and magnetic fields, pulsed and polarized radio and TV signals, microwaves, etc., i.e. environmental exposures that have never been around on our Planet, or – if they have – been less than one quintillionth in strength?"* And the answer is so simple (*i.e.* no "rocket science" needed!): Of course, we can not count on any such protective shielding since it is just not present. We are more naked than any newborn baby when it comes to such presumed protection.

In that chapter, I also concluded that once upon a time we all believed that radioactivity from radium, uranium and plutonium, the X-rays in medicine, as well as the ultraviolet sunrays, all were safe, as were very dangerous habits like smoking. We did not realize that they can harm us, indeed even kill us. In the 1940s kids' shoe shops were equipped with shoefitting machines that used strong X-rays, and wristwatches in the 1950s glowed in the dark because they were painted with radioactive paint. At the same time, responsible scientists and doctors started to realize that the warm and beautiful sunshine could actually can harm our cells and their DNA, leading to the development of skin cancer. The same sort of experts that today tell you that cell phone and Wi-Fi radiation is harmless, once told you that strong radioactivity, strong X-rays and UV light were harmless. And smoking, they said, even was good for you! Most of these hazards were quickly removed and are now gone, but a new one has appeared: The wireless society with all it's EMF-based gadgets.

Very early on I coined the expression and question that we are all subjected to *"The largest full-scale experiment ever: What happens when we, 24-hours around the clock, wherever we are, allow ourselves and our children to be used as guinea-pigs, whole-body-irradiated by new, man-made electromagnetic fields for the rest of our lives?"* This question is now more valid and important than ever and it is not a matter of "rocket science", the obvious evolutionary consequences are easy to grasp, and *it is time to wake up and take strong action! (As the 'major players', including the insurance and reinsurance industry, already have done more than 15 years ago!)*

Down below, I will bring up a few recent reports and results from the peer review-based scientific literature, acknowledged by other scientists in the field. Even though all the gadgets in question represent highly attractive technological developments claimed to improve our life, ease our everyday work, and amuse us, they do expose us to a potent toxic environmental pollution. The power-frequent electromagnetic fields and the microwave radiation may affect prenatal development in both humans and animals, as well as the health for children, teenagers, and adults. Various studies have reported that different types of artificial EMFs may have serious, or very serious, adverse side-effects in various organs, tissues, cells, and molecular classes, and especially so in the young and very young. It is very important that the members of the general public _immediately_ start to educate themselves and take precautionary actions of their own. I am constantly asking society to continue to build knowledge towards a safe future, built on what I have coined as «green, human- and environmental-friendly technology», a technology that should not be safer, but _safe!_ For instance, the use of a wireless connection is not necessary for access to the Internet. Hardwired Internet access using shielded cables and computers predated the use of wireless connection, and will still serve us equally well; the information will be exactly the same, the pictures and movies too, and we can still "like" each other at a distance.

It has been a great honour for me to assist Dr. Dimitris Panagopoulos and Dr. George Carlo, revealing the ground-breaking biomedical and biological importance of the polarization of artificial fields _versus_ the non-polarized character of natural fields; demonstrating the very limited use of the so-called SAR-values (=specific absorption rates) as dosimetric quantities for electromagnetic field bioeffects; and to point to the very basic, important and natural realization of always using real mobile phone exposures in experimental studies, instead of simulated ones (cf. Panagopoulos et al 2013, 2015a,b). Together we are now heading towards finding protective solutions, enabling consumers to continue using their various electronic gadgets, but in a safe way.

Among the most recent papers, several ought to attract strong attention, such as the one by Parsanezhad _et al_ (2017) where the health effects of mobile phone jammers - preventing the mobile phones from receiving signals from base stations by interfering with authorized mobile carriers' services – was studied. In spite of the fact that mobile jammer use most often is illegal, they are occasionally used in offices, shrines, conference rooms and cinemas. The purpose of this study was to investigate the biological effects of short-term exposure of human sperm to radiofrequency radiation emitted from a commercial mobile phone jammer.

Fresh semen samples were obtained by masturbation from 50 healthy donors who had been referred with their wives to the Infertility Treatment Center at the Mother and Child Hospital, Shiraz University of Medical Sciences, in Iran. Female problems were diagnosed as the reason for infertility in these couples. The semen sample of each participant was divided into 4 aliquots. The first aliquot was subjected to swim-up and exposed to jammer radiation. The second aliquot was not subjected to swim-up but was exposed to jammer radiation. The third and fourth aliquots were not exposed to jammer radiation but only the 3rd aliquot was subjected to swim-up.

The results revealed that the semen samples exposed to radiofrequency radiation showed a significant decrease in sperm motility and increase in DNA fragmentation, which lead to the authors' conclusion that electromagnetic radiation in the radiofrequency range emitted from mobile phone jammers may lead to decreased motility and increased DNA fragmentation in human semen. It can, thus, be concluded that – in addition to previous investigations using mobile phones only – also mobile phone jamming might exert adverse reproductive health effects.

Furthermore, Solek et al (2017) have investigated the effects of pulsed and continuous electromagnetic fields (PEMFs/CEMFs) on mouse spermatogenic cell lines (GC-1 spg and GC 2 spd) in terms of cellular and biochemical features in vitro. The authors evaluated the effect of EMFs on mitochondrial metabolism, morphology, proliferation rate, viability, cell cycle progression, oxidative stress balance and regulatory proteins. Their results strongly suggest that EMFs induce oxidative and nitrosative stress–mediated DNA damage, resulting in p53/p21-dependent cell cycle arrest and apoptosis.

Therefore, spermatogenic cells, due to the lack of antioxidant enzymes, undergo oxidative and nitrosative stress-mediated cytotoxic and genotoxic events, which contribute to infertility by reduction in healthy sperm cells pool. In conclusion, electromagnetic field present in surrounding environment impairs male fertility by inducing p53/p21-mediated cell cycle arrest and apoptosis.

Naturally, one should not forget other life-style factors which may affect us, and maybe in concert with EMFs. For instance Radwan *et al* (2016) found evidence for a relationship between sperm DNA damage parameters and everyday life factors. High and medium level of occupational stress and age increase DNA fragmentation index (p=0.03, p=0.004 and p=0.03, respectively). Other lifestyle factors that were positively associated with percentage of immature sperms (high DNA stainability index) included: obesity *and cell phone use for more than 10 years* (p=0.02 and p=0.04, respectively). Thus, data from the present study showed a significant effect of age, obesity, mobile phone use and occupational stress on sperm DNA damage. As DNA fragmentation represents an extremely important parameter indicative of infertility and potential outcome of assisted reproduction treatment, and most of the lifestyle factors are easily modifiable, the information about factors that may affect DNA damage are important, and should be reflected in precautionary societal advice to the general public.

Some of the first observations on human sperm cells were done by Agarwal *et al* (2008) who showed that the use of cell phones decrease the semen quality in men by decreasing the sperm count, motility, viability, and normal morphology. The decrease in sperm parameters was dependent on the duration of daily exposure to cell phones and independent of the initial semen quality.

To investigate the potential combined influence of maternal restraint stress and 2.45 GHz Wi-Fi signal exposure on postnatal development and behavior in the offspring of exposed rats, Othman *et al* (2017) studied 24 pregnant albino Wistar rats who were randomly assigned to four groups: Control, Wi-Fi-exposed, restrained and both Wi-Fi-exposed and restrained groups. Each of Wi-Fi exposure and restraint occurred 2 h/day along gestation till parturition. The pups were evaluated for physical development and neuromotor maturation. Moreover, elevated plus maze test, open field activity and stationary beam test were also determined on postnatal days 28, 30 and 31, respectively. After behavioral tests, the rats were anesthetized and their brains were removed for biochemical analysis. Their main findings showed no detrimental effects on gestation progress and outcomes at delivery in all groups. Subsequently, Wi-Fi and restraint, *per se* and mainly *in concert* altered physical development of pups with slight differences between genders. Behaviorally, the gestational Wi-Fi irradiation, restraint and especially the associated treatment affected the neuromotor maturation mainly in male progeny. At adult age, they noticed anxiety, motor deficit and exploratory behavior impairment in male offspring co-exposed to Wi-Fi radiation and restraint, and in female progeny subjected to three treatments. The biochemical investigation showed that, all three treatments produced global oxidative stress in brain of both sexes. As for serum biochemistry, phosphorus, magnesium, glucose, triglycerides and calcium levels were disrupted. Taken together, prenatal Wi-Fi radiation and restraint, alone and combined, provoked several behavioral and biochemical impairments at both juvenile and adult age of the offspring.

Hassanshahi *et al* (2017) aimed to investigate the effect of 2.4 GHz Wi-Fi radiation on multisensory integration in rats. This experimental study was done on 80 male Wistar rats that were allocated into exposure and sham groups. Wi-Fi exposure to 2.4 GHz microwaves [in Service Set Identifier mode (23.6 dBm and 3% for power and duty cycle, respectively)] was done for 30 days (12 h/ day). Cross-modal visual-tactile object recognition (CMOR) task was performed by four variations of spontaneous object recognition (SOR) test including standard SOR, tactile SOR, visual SOR, and CMOR tests. A discrimination ratio was calculated to assess the preference of animal to the novel object. The expression levels of M1 and GAT1 mRNA in the hippocampus were assessed by quantitative real-time (RT) PCR. Results demonstrated that rats in Wi-Fi exposure groups could not discriminate significantly between the novel and familiar objects in any of the standard SOR, tactile SOR, visual SOR, and CMOR tests. The expression of M1 receptors increased following Wi-Fi exposure. In conclusion, results of this study showed that chronic exposure to Wi-Fi electromagnetic waves might impair both unimodal and cross-modal encoding of information.

Rezk *et al* (2008) provided evidence that exposure of pregnant women to mobile phone significantly increase fetal and neonatal heart rate, and significantly decreased the cardiac output, and Lai *et al* (1994) demonstrated that after 45 min of exposure to pulsed 2,450 MHz microwaves (2 microseconds pulses, 500 pps, 1 mW/cm2, average whole body SAR 0.6 W/kg), rats showed retarded learning while performing in the radial-arm maze to obtain food rewards, indicating a deficit in spatial "working memory" function. Their data indicate that both cholinergic and endogenous opioid neurotransmitter systems in the brain are involved in the microwave-induced spatial memory deficit. Highly similar conclusions were reached by Papageorgiou *et al* (2011) through their findings in young men suggesting that Wi-Fi exposure may exert gender-related alterations on neural activity associated with the amount of attentional resources engaged during a linguistic test adjusted to induce working memory.

In a very interesting study, Cervellati *et al* (2009) were able to demonstrate a significant effect of high-frequency electromagnetic fields on connexins expression and localization in placental extravillous trophoblast cell line HTR-8/SVneo (trophoblasts are cells forming the outer layer of a blastocyst, which provide nutrients to the embryo and develop into a large part of the placenta). Connexins are membrane proteins able to influence trophoblast functions. Samples were exposed to pulse-modulated 1817 MHz sinusoidal waves (GSM-217 Hz; 1 h: SAR of 2 W/kg [=the maximal allowed public exposure level]). Connexin mRNA expression was assessed through semi-quantitative RT-PCR, protein expression by Western blotting, protein localization by indirect immunoflorescence, cell ultrastructure using electron microscopy. The exposure significantly and selectively increased Cx40 and Cx43, without altering protein expression. Nevertheless, Cx40 and Cx43 lost their punctuate fluorescence within the cell membrane, becoming diffuse after HF-EMF exposure. Electron microscopy evidenced a sharp decrease in intercellular gap junction-like structures.

This study is the first to indicate that exposure of extravillous trophoblast to GSM signals can modify connexin gene expression, connexin protein localization and cellular ultrastructure, and they may also explain – together with the above-mentioned studies on sperm cells - why infertility was encountered in the Greek study by Magras & Xenos (1997), where a progressive decrease in the number of newborns per dam was observed, which ended in irreversible infertility, after in vivo exposures at several places around an antenna park outside of the city of Thessaloniki. At these locations, the radiofrequency power density was between 1,680 $\mu W/m^2$ and 10,530 $\mu W/m^2$, the latter being a typical exposure value 100 meters from a base station/antenna. The prenatal development of the newborns, however, evaluated by the crown-rump length, the body weight, and the number of the lumbar, sacral, and coccygeal vertebrae, was improved, something which may sound appetizing. But, remember, any abnormal pattern must always be regarded as just that: abnormal. To feed these fetuses with energy may have 'developed' them – just as feeding a body-builder with anabolic steroids, but the latter then will get a dramatic reduction is genital development and fertility scores, just as the mice outside of Thessaloniki did. (Ask any professional body-builder if you do not believe me. Or ask a professional gardener what happens I you feed blooming plants with way too much fertilizers (=energy); they will get huge green masses but very few and tiny flowers, if any. It is as simple as that, it is my working hypothesis, and you should quote it and demand research into this area of mechanistic approach.)

Cerón-Carrasco & Jacquemin (2017), in their elegant study, have pointed to the fact that nowadays, using e.g. mobile phones as electromagnetic 'knives', employing their pulsed electric fields to selectively rewrite the stored genetic information. However, for such modification to be effective, one needs, as a prerequisite, that the replication mechanism is not stopped by the field, so that the changes propagate over the following generations. Cerón-Carrasco & Jacquemin (2017) used theoretical calculations to demonstrate that while such fields lead to permanent noncanonical Watson-Crick guanine-cytosine (GC) base pairs, the G-quadruplex motifs present in telomeres can more effectively preserve their native forms. Indeed, G-quadruplexes "resist" the perturbations induced by field strengths going up to 60×10-4 a.u., a figure constituting the upper limit before the complete destruction of the double helix architecture. As the authors point out, since the induced errors in the DNA base pairs are not transcribed into the telomeres, electric fields can indeed be used as a source of selective mutations in the genetic code.

To protect their core machinery from the attack of exogenous agents, cells locate DNA in their nucleus (and in their mitochondria). Nevertheless, some reactive chemical species and physical agents might reach DNA and alter its natural double helix structure. Such interactions may be used in a laboratory setting to non-invasively alter the genetic make-up, but − if let loose in society! − may harm us all, something already shown several times over the last decades.

In addition to the above, Sun *et al* (2017) employed HL-60 cells, derived from human promyelocytic leukemia, and exposed them to continuous wave 900 MHz radiofrequency fields (RF) at 120 µW/cm2 power intensity for 4h/day for 5 consecutive days to examine whether such exposure is capable of damaging the mitochondrial DNA (mtDNA) mediated through the production of reactive oxygen species (ROS). In addition, the effect of RF exposure was examined on 8-hydroxy-2'-dexoyguanosine (8-OHdG) which is a biomarker for oxidative damage and on the mitochondrial synthesis of adenosine triphosphate (ATP) which is the energy required for cellular functions. The results indicated a significant increase in ROS and significant decreases in mitochondrial transcription factor A, mtDNA polymerase gamma, mtDNA transcripts and mtDNA copy number in RF-exposed cells compared with those in sham-exposed control cells. In addition, there was a significant increase in 8-OHdG and a significant decrease in ATP in RF-exposed cells. The response in positive control cells exposed to gamma radiation (GR, which is also known to induce ROS) was similar to those in RF-exposed cells. Thus, the overall data indicated that RF exposure was capable of inducing mtDNA damage mediated through ROS pathway which also induced oxidative damage. Very interestingly, prior-treatment of RF- and GR-exposed the cells with melatonin, a well-known free radical scavenger, reversed the effects observed in RF-exposed cells.

Mounting evidence suggests that exposure to radiofrequency electromagnetic radiation (RF-EMR) can influence learning and memory in rodents, primarily reducing the concentration capacity and short-term memory. However, in a recent study by Wang *et al* (2017), they examined the supportive effects of single exposure to 1.8 GHz RF-EMR for 30 min on subsequent recognition memory in mice, using the novel object recognition task (NORT). RF-EMR exposure at an intensity of >2.2 W/kg specific absorption rate (SAR) induced a significant density-dependent increase in NORT index with no corresponding changes in spontaneous locomotor activity. RF-EMR exposure increased dendritic-spine density and length in hippocampal and prefrontal cortical neurons, as shown by Golgi staining. Whole-cell recordings in acute hippocampal and medial prefrontal cortical slices showed that RF-EMR exposure significantly altered the resting membrane potential and action potential frequency, and reduced the action potential half-width, threshold, and onset delay in pyramidal neurons. These results demonstrate that exposure to 1.8 GHz RF-EMR for only 30 min (!) can significantly increase recognition memory in mice, and can change dendritic-spine morphology and neuronal excitability in the hippocampus and prefrontal cortex. The SAR in this study (3.3 W/kg) was slightly outside the range mostly encountered in normal daily life, and its relevance as a potential therapeutic approach for disorders associated with recognition memory deficits naturally remains to be clarified (also see Panagopoulos *et al* 2013). As already stated above, according to my own experience, it should also be remembered that even if an effect might seem nice, positive, and tempting to employ for treatment, such effects should always be cautioned as abnormal, and it might be much more correct to deal with the underlying casue rather than artificially treat unwanted outcomes as they appear. In the above case it would seem very good to allow for a significant density-dependent increase in the NORT index, but such reasoning could easily then lead people to naively similarly believe in, and urge for, mind-altering drugs and other pharmaceutical agents. Our, and the rodents', learning and memory capacity is quite enough without any outside or inside manipulation.

In contrast to this seemingly positive effect, Erkut *et al* (2016) investigated the effects of exposure to an 1,800 MHz electromagnetic field on bone development during the prenatal period in rats, and found that increasing the duration of exposure during the prenatal period resulted in a significant reduction of resting cartilage levels and a significant increase in the number of apoptotic chondrocytes and myocytes. There was also a reduction in calcineurin activities in both bone and muscle tissues. They observed that the development of the femur, tibia, and ulna were negatively affected, especially with a daily EMF exposure of 24 hours. So, in essence, bone and muscle tissue development was negatively affected due to prenatal exposure to an 1,800 MHz radiofrequency electromagnetic field.

Taheri *et al* (2017) assessed if the exposure to 900 MHz GSM mobile phone radiation and 2.4 GHz radiofrequency radiation emitted from common Wi-Fi routers alters the susceptibility of microorganisms to different antibiotics. Pure cultures of *Listeria monocytogenes* and *Escherichia coli* were exposed to RF-EMFs generated either by a GSM 900 MHz mobile phone simulator or a common 2.4 GHz Wi-Fi router. It was shown that exposure to RF-EMFs within a narrow level of irradiation (an exposure window) makes microorganisms resistant to antibiotics. Altogether, the findings of this study showed that exposure to Wi-Fi and RF simulator radiation can significantly alter the inhibition zone diameters and growth rate for *L monocytogenes* and *E coli*. These findings may have implications for the management of serious infectious diseases. *With the on-going huge and highly frightening development into more and more antibiotics-resistant microorganisms around the world, this adaptive phenomenon and its potential threats to human health, according to my view, definitely and rapidly should be further investigated in future experiments!* At this point in time, to instead disengage academic scientists from their workplaces due to "lack of money" will not sound well in the future.

There is also emerging evidence that wireless, non-ionizing radiation (from cell phones, Wi-Fi, and smart meters) harms wildlife and damages trees. There have been direct reports of such radiation affecting vital bee populations (cf. above), disturbing bird habitats, and interfering with avian navigational systems. French researchers, under the direction of Alain Vian at the Equipe de Recherche Transduction et Autosurveillance Cellulaire, Universite Blaise Pascal in Aubière, have shown that even tomato plants react to the damage from the relatively weak 900 MHz radiation from cell towers (Roux *et al* 2008). The scientists believe they found an environmental factor that instantly impacts the genetic material in the tomato cells, which in turn resulted in the tomato plant cells reacting with a chemical damage sequence, involving the molecule calmodulin. The effect was described in public interviews as "exactly as if we had crushed them with a hammer," by the scientists.

It was enough to expose a few leaves of the plant for the entire plant to react. The damage was lessened, however, on the parts of the plant that were shielded from the radiation.

The interesting thing about tomatoes is that they can not cheat or be swayed by emotions or expectations

They have no conscience.

They can not move.

They do not cheat the insurance company for money.

They are not imagining things.

The don't blame their workplace problems on alleged "electrical over-sensitivity."

They don't read newspapers, listen to radio, or watch TV (so they can't fall victim for any massmedia-driven psychosis).

They are instead very sensitive to their surrounding environment and are fussy when it comes to conditions for their survival.

Had the French tomato plants been able to escape, they obviously would have done so.

Finally, in a replication study, following the preliminary findings of five Danish schoolgirls, we studied the effect of mobile phone base station signals on common *Brassicaceae Lepidium sativum* (cress d'Alinois) seed germination (Cammaerts & Johansson 2015). Under high levels of radiation (70-100 μW/m2 =175 mV/m), the seeds never germinated. In fact, the first step of the seeds' germination – the imbibitions of germinal cells – could not occur under radiation, while inside the humid compost such imbibitions occurred and roots slightly developed. When removed from the electromagnetic field, seeds germinated normally. The radiation was, thus, most likely the cause of the non–occurrence of the seeds' imbibitions and germination.

In conclusion, the present investigation - although preliminary in its character - indicates that the prodigious wireless technology may effectively and seriously impact nature and should urgently be used much more cautiously, or maybe even not at all. The present study also brings some new information on the subject - effect of electromagnetism on plants - but it must be replicated on several plants species, at different independent laboratories, as well as developed further at the cytological and physiological levels by botanists, histologists and physiologists. Finally, in essence, it clearly supports the initial findings of Lea Nielson, Mathilde Nielsen, Signe Nielsen, Sisse Coltau and Rikke Holm, at Hjallerup Skole, under the supervision of their biology teacher Mr. Kim Horsevad.

Society of today employs a variety of wireless technologies using transmitters that emit electromagnetic waves creating radiation and electromagnetic fields. The research covered above – together with a huge number of other reports – clearly demonstrate that, at the power levels required for these wireless technologies to operate reliably, the radiofrequency radiation as well as the low-frequency fields have significant biological and biomedical effects, many of which - from a human perspective - must be considered as very serious and alarming. Thus, a rapidly accumulating body of scientific evidence of harm to health and well-being constitute early warnings that adverse health effects can occur with prolonged exposures to very "low-intensity" (remember again that the exposure levels that are regarded as "low-intensity" actually are astronomically high compared to natural background levels) electromagnetic fields at biologically active frequencies or frequency combinations/windows. The consequences of such exposures can be especially grave for electrohypersensitive individuals and children. The telecom industry uses inapplicable health safety standards, which I have pointed to above, and flawed reasoning to promote the safety of their products, eagerly backed by a naive and uninformed political establishment. However, in contrast to this, because the effects are reproducibly observed, and links to pathology can not be excluded, the Precautionary Principle – or a complete ban! - should be in force regarding the implementation of these new technologies within society.

From the current vast scientific literature, it is obvious we must proceed with caution before immersing the citizens in more and more artificial electromagnetic fields. We may, as a matter of fact, already be gravely endangering our current as well as coming generations. To not act today, may prove a disaster tomorrow, and such lack of action may again result in the classical "*late lessons from early warnings*".

I, as a scientist, is not here to promote convenience or economic growth, but only "*to protect and serve*" human health and biological safety, as well as to protect other animals, plants, and bacteria. These aims must be our only target.

In November, 2009, a Scientific Panel comprised of international experts on the biological effects of electromagnetic fields met in Seletun, Norway, for three days of intensive discussion on existing scientific evidence and public health implications of the unprecedented global exposures to artificial electromagnetic fields from telecommunications and electric power technologies, and ending in The Seletun Scientific Statement (Fragopoulou *et al* 2010), which strongly recommends that lower limits (<0.017 µW/cm2) be established for electromagnetic fields and wireless exposures. This meeting was a direct consequence of on-going discussions since the mid-nineties, when cellular communications infrastructure began to rapidly proliferate. At the beginning of the 21st century, many resolutions, like the Benevento (Belpoggi *et al* 2006), Venice (Avino *et al* 2008) and London (Johansson 2009b) Resolutions were created to protect health. Important conclusions were drawn from the 600-page Bioinitiative Report [http://www.bioinitiative.org] published August 31, 2007, which was a review of over 2,000 studies showing biological effects from electromagnetic radiation at non-thermal levels of exposure, and which later was partly published in the medical journal Pathophysiology (Volume 16, 2009). The Bioinitiative Report has since been updated (2012; 2014).

Many researchers now believe the existing safety limits are inadequate to protect public health because they do not consider prolonged exposure to lower emission levels that are now widespread and do not take into account non-thermal effects. It should be noted that only one hygienic safety value ever has been proposed: 0.0000000001-0.000000000000001 µW/cm2 – this is the natural background during normal cosmic activities; proposed by·myself at a trade union meeting in Stockholm, already in 1997, as a genuine hygienic safety value, and since then many times repeatedly presented. (Given the highly artificial nature of the current wireless communication signals, e.g. of their pulsations and modulations, it may actually boil down to 0 (zero) µW/cm2 as the true safe level.) And do not ever believe it is possible to play it "safer" by only somewhat reducing the exposure levels!

The conversion to Wi-Fi, and similar wireless communication systems, is one of the largest technology rollouts in history, and yet virtually no public consultation with citizens or local governments was carried out in advance. Parallel to this, the World Health Organization (WHO; May 31, 2011) has classified the radiofrequency radiation used as a possible carcinogen, and the world's insurance companies have abandoned ship by not insuring or reinsuring for health effects of electromagnetic fields. Around the world Wi-Fi companies continue to install their antennas, often without public awareness or consent, and now in an ever-accelerating roll-out pace with 5G and the "Internet of Things". According to my view, this is a genuine threat to our democracy and informed decision-making, and it is definitely fair to call for immediate and strong precautionary measures as well as much better monitoring of health parameters changes in our modern societies (cf. Hallberg & Johansson 2009).

I have many times written about the human and environmental rights aspects in various debate articles and commentaries*, and pointed to them in many of my public lectures, and in radio and TV interviews. As you already understand, they are very important for us and life on the Planet, and ought to be properly addressed. For such addresses to be successful, you do need the right advocates on your side, but also the right media, I would say. So much nowadays is fought both in courts of law as well as on the newspaper pages, and in radio and TV.

*It's last paragraph reads:

"I samband med rättegångarna i Nürnberg efter andra världskrigets slut formulerades för första gången en offentlig etisk kod för medicinska experiment som involverar människan, Nürnbergkodexen 1947. Här slogs bl.a. fast att informerat samtycke krävs och att riskerna för försökspersoner skall minimeras. Det framhölls att varje deltagare har rätt att när som helst avbryta sitt deltagande i ett experiment och att den som leder ett sådant skall avbryta det

---

* See e.g. Johansson O, Gullbrandsson A, Dämvik M, Hallberg Ö, Hellberg K, Lindkvist L, "Ohälsan ökar i takt med strålningen" (= "Ill health increases with radiation", in Swedish), Borås Tidning 14/2 2011.

om det verkar troligt att en deltagare skadas. När får vi avbryta vårt deltagande i det pågående strålningsexperimentet?"

English translation using Google Translator:

"In connection with the Nuremberg trials after the Second World War it was formulated for the first time a public code of ethics for medical experiments involving human beings, the Nuremberg Code of 1947. Among other things stated was that informed consent is required and that the risks to subjects are minimized. It was pointed out that each participant has the right at any time to cancel their participation in an experiment, and that whoever leads such should cancel it if it seems likely that a participant is injured. When may we suspend our participation in the ongoing radiation experiment?"

In Sweden we often say that "people are trying to invent the wheel again". Don't, I say. Use instead common sense, the knowledge and the tools that are already in front of you, and be precise. Demand only what you want, nothing more, nothing less. Be clear. Be bold. NEVER give up.

I just wish I could have done more for you, and for life on our Planet. Both in science as well as in politics (*sic!*).

In summary: Do not believe that mobile phones, iPads and Wi-Fi are safe; they are not! (And the major 'players' in our society know it.) *They interfere with normal brain function, learning and memory, fertility, cancer risks and have been shown to shatter the DNA in cells. All of this can be found in peer-reviewed scientific journals but, until now, has not been in the public domain. But very few will try to protect you and very few want to speak the truth.* (Does this sound good to you?) *So maybe the only correct answer to my question above is: No more full-scale experimentation until all the 'major players' climb on board again to cover any form of future legal liability claims?!*

With all the new data from different investigations appearing, some days with several publications being released, maybe I was not wrong when I called for safety measures already back in the early 1980ies; maybe it was morally-ethically 100% right to sound the alarm? Along these lines I could not help smiling when I read, in June 2018, in Dagens Nyheter, one of the largest daily newspapers in Sweden, that a Karolinska Institute-based colleague of mine now indicated that human sperm reduction might be due to several factors, including cell phone and computer radiation (https://www.dn.se/insidan/halvering-av-mannens-spermier-oroar/). This very same professor refused to collaborate around this issue 20 years ago when I approached him with the very same hypothesis. So, as usual, time changes our perspectives.

Finally, I say, from a public health point-of-view no more research is needed, the proof in the form of thousands and thousands of peer review-based scientific publications is overwhelming – now society must dare to protect and to serve. Children can never be allowed to be victims of flimsy pedagogic tools, and absent adult responsibility, or to be exposed to a WHO-classified possible carcinogen. Our actions must solely aim for their needs, not for commercial greed.

Personally, I would hate to arrive at the Pearly Gates and hear Saint Peter say: *"Why did you not react and act, Olle, you understood, you knew, you saw; you could and should have done much more!"*. No, as a mental fire brigade soldier, I rather try my hardest and possibly be wrong – false alarms never make the ordinary fire fighters or citizens weep, and so it should not make anyone sad or angry if my concern is wrong. (...But if I am right, then what...?)

ACKNOWLEDGEMENTS

This paper is dedicated to all those brave persons who dared to use their common sense, and who dared to speak out, and stand up. I also want to thank all of you – and you know who you are! – for your mental, practical, inspirational, and economical support.

This specific paper was supported by the Karolinska Institute, and a grant from Mr. Einar Rasmussen, Kristiansand S, Norway. Mr. Brian Stein, Melton Mowbray, Leicestershire, UK, Dr. Toril H. Jelter, Mount Diablo Integrated Wellness Center, Walnut Creek,

California, USA, and the Irish Doctors Environmental Association (IDEA; Cumann Comhshaoil Dhoctúirí na hÉireann) are gratefully acknowledged for their very important contributions.

REFERENCES

Agarwal A, Deepinder F, Sharma RK, Ranga G, Li J, "Effect of cell phone usage on semen analysis in men attending infertility clinic: an observational study", Fertil Steril 2008; 89: 124-128

Avino P, d'Alessandro A, Bedini A, Belyaev I, Belpoggi F, Blackman C, Blank M, Bobkova N, Bruno B, Cinti C, Cristaldi M, Dasdag S, De Ninno A, Del Giudice E, de Salles A, Doull S, Georgiou C, Goodman R, Grimaldi S, Giuliani L, Hardell L, Havas M, Hyland G, Lisi A, Ieradi L, Johansson O, Khurana VG, Lai H, Margaritas L, Marinelli F, Markovic V, Maxey E, Oberfeld G, Phillips J, Richter E, Salford L, Scalia M, Seyhan N, Shalita Z, Soffritti M, Szmigielski S, Udroiu I, Verduccio C, Zeyrek M, Zhadin M, Zinelis S, Zucchero A, Goldsworthy A, "The Venice Resolution 2008",http://www.icems.eu/resolution.htm

Belpoggi F, Blackman CF, Blank M, Bobkova N, Boella F, Cao Z, Allessandro AD, Emilia ED, Del Giuduice E, De Ninno A, De Salles AA, Giuliani L, Grigoryev Y, Grimaldi S, Hardell L, Havas M, Hyland G, Johansson O, Kundi M, Lai HC, Ledda M, Lin Y-P, Lisi A, Marinelli F, Richter E, Rosola E, Salford L, Seyhan N, Soffritti M, Ramazzini B, Szmigielski S, Zhadin M, "Benevento Resolution 2006", Electromag Biol Med 2006; 25: 197–200

Cammaerts M-C, "Is electromagnetism one of the causes of the CCD? A work plan for testing this hypothesis", J Behav 2017; 2: 1-6

Cammaerts MC, Johansson O, "Ants can be used as bio-indicators to reveal biological effects of electromagnetic waves from some wireless apparatus", Electromagn Biol Med 2013; 33: 282-288

Cammaerts MC, Johansson O, "Effect of man-made electromagnetic fields on common *Brassicaceae Lepidium sativum* (cress d'Alinois) seed germination: a preliminary replication study", Phyton, International Journal of Experimental Botany 2015; 84: 132-137

Cerón-Carrasco JP, Jacquemin D, "Exposing the G-quadruplex to electric fields: the role played by telomeres in the propagation of DNA errors", Phys Chem Chem Phys 2017 Mar 20; doi: 10.1039/c7cp01034f. [Epub ahead of print]

Cervellati F, Franceschetti G, Lunghi L, Franzellitti S, Valbonesi P, Fabbri E, Biondi C, Vesce F, "Effect of high-frequency electromagnetic fields on trophoblastic connexins", Reprod Toxicol 2009; 28: 59-65

Dämvik M, Johansson O, "Health risk assessment of electromagnetic fields: A conflict between the precautionary principle and environmental medicine methodology", Rev Environ Health 2010; 25: 325-333

Erkut A, Tumkaya L, Balik MS , Kalkan Y, Guvercin Y, Yilmaz A, Yuce S, Erkan Cure E, Sehitoglu I, "The effect of prenatal exposure to 1800 MHz electromagnetic field on calcineurin and bone development in rats", Acta Cir Bras 2016; 31: http://dx.doi.org/10.1590/S0102-865020160020000001

Falcioni L, Bua L, Tibaldi E, Lauriola M, De Angelis L, Gnudi F, Mandrioli D, Manservigi M, Manservisi F, Manzoli I, Menghetti I, Montella R, Panzacchi S, Sgargi D, Strollo V, Vornoli A, Belpoggi F, "Report of final results regarding brain and heart tumors in Sprague-Dawley rats exposed from prenatal life until natural death to mobile phone radiofrequency field representative of a 1.8 GHz GSM base station environmental emission", Environ Res 2018, 165: 496-503

Fragopoulou A, Grigoriev Y, Johansson O, Margaritis LH, Morgan L, Richter E, Sage C, "Scientific panel on electromagnetic field health risks: Consensus points, recommendations, and rationales. Scientific Meeting: Seletun, Norway, November 17-21, 2009", Rev Environ Health 2010; 25: 307-317

French PW, Penny R, Laurence JA, McKenzie DR, "Mobile phones, heat shock proteins and cancer", Differentiation 2001; 67: 93–97

Hallberg Ö, Johansson O, "Apparent decreases in Swedish public health indicators after 1997 — Are they due to improved diagnostics or to environmental factors?", Pathophysiology 2009; 16: 43–46

Hardell L, Carlberg M, Söderqvist F, Mild KH, "Long-term use of cellular phones and brain tumours: increased risk associated with use for ≥10 years", Occup Environ Med 2007; 64: 626–632

Hassanshahi A, Shafeie SA, Fatemi I, Hassanshahi E, Allahtavakoli M, Shabani M, Roohbakhsh A, Shamsizadeh A, "The effect of Wi-Fi electromagnetic waves in unimodal and multimodal object recognition tasks in male rats", Neurol Sci 2017 Mar 22 [Epub ahead of print]

Herbert MR, Sage C, "Autism and EMF? Plausibility of a pathophysiological link - Part I", Pathophysiology 2013a; 20: 191-209.

Herbert MR, Sage C, "Autism and EMF? Plausibility of a pathophysiological link - Part II", Pathophysiology. 2013b; 20: 211-34

Johansson O, "Disturbance of the immune system by electromagnetic fields — A potentially underlying cause for cellular damage and tissue repair reduction which could lead to disease and impairment", Pathophysiology 2009a; 16: 157-177

Johansson O, "The London Resolution", Pathophysiology 2009b; 16: 247-248

Johansson O, "Electrohypersensitivity: a functional impairment due to an inaccessible environment", Rev Environ Health 2015; 30: 311–321

Johansson O, "Health effects of artificial electromagnetic fields: A wake-up call from a neuroscientist... But is anyone in power picking up? Hello...?", In: 2016 Environmental Sensitivities Symposium: TextBook (ed. L Curran), Building Vitality, Carlton North, 2016, pp 73-94, ISBN 13:978-1539094227

Lai H, Horita A, Guy AW, "Microwave irradiation affects radial-arm maze performance in the rat", Bioelectromagnetics 1994; 15: 95-104

Magras IN, Xenos TD, "RF radiation-induced changes in the prenatal development of mice", Bioelectromagnetics 1997; 18: 455-461

Mariea TJ, Carlo GL, "Wireless radiation in the etiology and treatment of autism: clinical observations and mechanisms", J Aust Coll Nutr & Env Med 2007; 26: 3-7

Othman H, Ammari M, Sakly M, Abdelmelek H, "Effects of prenatal exposure to WIFI signal (2.45 GHz) on postnatal development and behavior in rat: Influence of maternal restraint", Behavioural Brain Research 2017; 36: 291-302

Panagopoulos DJ, Johansson O, Carlo GL, "Evaluation of specific absorption rate as a dosimetric quantity for electromagnetic fields bioeffects", PLoS ONE 2013; 8: e62663. doi:10.1371/journal.pone.0062663

Panagopoulos DJ, Johansson O, Carlo GL, "Real versus simulated mobile phone exposures in experimental studies", BioMed Res Internat 2015a, Article ID 607053, http://dx.doi.org/10.1155/2015/607053

Panagopoulos DJ, Johansson O, Carlo GL, "Polarization: A key difference between man-made and natural electromagnetic fields, in regard to biological activity", Nature Scientific Reports 2015b; 5: 14914, 1-10, doi: 10.1038/srep14914

Papageorgiou CC, Hountala CD, Maganioti AE, Kyprianou MA, Rabavilas AD, Papadimitriou GN, Capsalis CN, "Effects of Wi-Fi signals on the P300 component of event-related potentials during an auditory Hayling task", J Integr Neurosci 2011, 10. 189-202

Parsanezhad ME, Mortazavi SMJ, Doohandeh T, Namavar Jahromil B, Mozdarani H, Zarei A, Davari M, Amjadi S, Soleimani A, Haghani M, "Exposure to radiofrequency radiation emitted from mobile phone jammers adversely affects the quality of human sperm", Int J Radiat Res 2017; 15 [Epub ahead of print]

Radwan M, Jurewicz J, Merecz-Kot D, Sobala W, Radwan P, Bochenek M, Hanke W, "Sperm DNA damage—the effect of stress and everyday life factors", Int J Impot Res 2016; Advance online publication, doi: 10.1038/ijir.2016.15

Rezk AY, Abdulqawi K, Mustafa RM, Abo El-Azm TM, Al-Inany H, "Fetal and neonatal responses following maternal exposure to mobile phones", Saudi Med J 2008; 29: 218-223

Roux D, Vian A, Girard S, Bonnet P, Paladian F, Davies E, G Ledoigt, "High frequency (900 MHz) low amplitude (5 V/m) electromagnetic field: a genuine environmental stimulus that affects transcription, translation, calcium and energy charge in tomato", Planta 2008; 227: 883-891

Solek P, Majchrowicz L, Bloniarz D, Krotoszynska E, Koziorowski M, "Pulsed or continuous electromagnetic field induce p53/p21-mediated apoptotic signaling pathway in mouse spermatogenic cells in vitro and thus may affect male fertility", Toxicology 2017 Mar 16; pii: S0300-483X(17)30092-6. doi: 10.1016/j.tox.2017.03.015. [Epub ahead of print]

Sun Y, Zong L, Gao Z, Zhu S, Tong J, Cao Y, "Mitochondrial DNA damage and oxidative damage in HL-60 cells exposed to 900MHz radiofrequency fields", Mutat Res 2017 Mar 7;797-799:7-14. doi: 10.1016/j.mrfmmm.2017.03.001

Taheri M, Mortazavi SMJ, Moradi M, Mansouri S, Hatam GR, Nouri F, "Evaluation of the effect of radiofrequency radiation emitted from Wi-Fi router and mobile phone simulator on the antibacterial susceptibility of pathogenic bacteria Listeria monocytogenes and Escherichia coli", Dose Response 2017; 15: 1559325816688527. Published online 2017 Jan 23. doi: 10.1177/1559325816688527

Wang K, Lu JM, Xing ZH, Zhao QR, Hu LQ, Xue L, Zhang J, Mei YA, "Effect of 1.8 GHz radiofrequency electromagnetic radiation on novel object associative recognition memory in mice", Sci Rep 2017; 7: 44521. doi: 10.1038/srep44521

SAVING 'FREE WILL' FROM SCIENCE

(Based on a true journey with my graduate advisor, Bill Banks)

Eve Isham

Ideas should not be judged scientific or unscientific, true or false, on the basis of their origins. Truths may come from sources that are quite unreliable, and false theories may come from the most trustworthy persons applying the most rigorous methods.

— *Paul D. Allison*

FIRST, LET ME INTRODUCE myself. I am a young experimental psychologist. I must emphasize the two adjectives "young" and "experimental." Young – meaning I am only starting out in my career and I am still sorting things out in my head. But this also means that I am open to new ideas instead of being married to a theory that desperately begs for a re-evaluation. "Experimental" reflects my scientific training and my appreciation for the value of careful experimental designs. I value conservative conclusions and scientific approaches that require the exhausting of all possibilities before claiming causation over correlation. Having said that, here is my story about how my graduate advisor, Bill Banks, and I tried to save free will.

REMINISCING ABOUT OUR FIRST STUDY

Benjamin Libet has become a household name for many philosophers, psychologists, and cognitive neuroscientists, and to those who are curious about free will. One of the extreme interpretations from Libet's hallmark study on the timing of action-related mental and neural events challenged the philosophical concept of free will (Libet, Gleason, Wright, & Pearl, 1983). Libet's innovative research in the early 1980's involved having the participants perform a simple voluntary action such as flexing the wrist. This action was described as voluntary and unplanned. While doing so, the participants observed a moving dot that rotated on the circular clock face (known as the Wundt clock after Wilhelm Wundt; Figure 1). When the flexion was executed, the participants referred to the location of the dot on the clock face to report the time they first experienced the urge to flex. This reported time is known as Libet's W. [this paragraph sounds boring compared to the earlier paragraph...I think you could reorder it to make it more exciting, something like: "Benjamin Libet is a house-hold name for many philosophers, psychologists and neuroscientists who wish to believe that the brain is deterministic and we do not have free-will." I would rephrase some of the sentences here to spell out their importance]

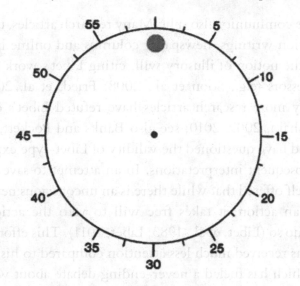

Fig. 1 Eve Isham

Depiction of the Wundt clock as used by Libet et al. The clock has a red spot that rotates clockwise and completes its rotation in 2.5 seconds.

The critical piece of how Libet's results can be interpreted to deny free will is the temporal relationship between the perceived time of urge or will ("W") as read from the clock and the onset time of the readiness potential (RP). Specifically, W is the the time of "will" of an action and the RP is the brain activity related to motion preparation (Kornhuber & Deecke, 1965). In Libet's case, he observed that the RP began significantly earlier than the moment of W, which he referred to as the conscious urge to act. Critically, this time lag of W led to the conclusion that the brain had already engaged in preparing for an action before the mind becomes aware of the intent to act. Thus, the philosophical perspective that conscious intent causes action became threatened, and the idea that we have the freedom to cause action is illusory.

As you might imagine, the results gained great popularity and even greater consequences. Determinists promoted the research findings extensively while dualists questioned the work even more.

The science community also split. Many research articles, textbooks, science fiction writings, newspaper columns and online blogs have embraced the notion of illusory will, citing Libet's work and work of its successors (e.g., Soon et al., 2008; Fried, et al. 2011). Vice versa, many more research articles have refuted Libet's data (e.g., Banks & Isham, 2009, 2010; see also Banks and Pockett, 2007 for review), and have questioned the validity of Libet-type experiments and the subsequent interpretations. In an attempt to save free will, Libet himself offered that while there is an unconscious neural event preceding an action, it takes free will to veto the action if one chooses to do so (Libet, et al., 1983; Libet, 2011). This effort by Libet however has received much less attention compared to his signature findings which has fueled a never-ending debate about whether or not we have free will – going strong in its third decade.

I was first introduced to this provoking idea when I was a research assistant in a lab at UCLA. The principal investigator was searching for the neural correlates of consciousness and had wanted to explore Libet's findings at the single cell levels. Bright-eyed and bushy tailed, I thought this was a cool idea, and became involved in the initial developmental stage of the work. However, my plans changed before I could complete the project. I entered a graduate program in eyewitness cognition at the Claremont Colleges in Southern California, leaving behind the Libet paradigm. During my first year doing eyewitness research, I met Bill Banks. Bill's passion for science was contagious, and along with his quirkiness and brilliant mind, my research interest again was redirected. I left eyewitness research to be trained with Bill. And as if it was pre-determined (pun intended), Bill led me back to the Libet paradigm.

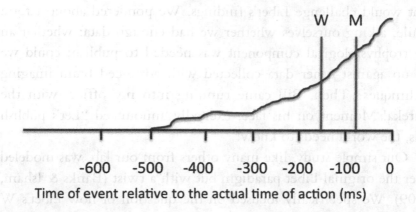

Fig. 2 Eve Isham

A depiction of Libet et al. (1983) findings. The readiness potential onset at approximately –500 to –1000 ms before the actual time of action. This onset precedes the perceived moments of urge (W) or of action (M) by 300 ms or more.

As the founder and the then editor-in-chief of the journal Consciousness and Cognition, Bill Banks was well submerged in the literature and the controversy revolving around Libet's findings. We began brainstorming about the responses, the reactions, and the criticisms, along with the things that we could do as scientists to validate (or invalidate) Libet's results. Although our bucket list was a long one, on top of it was an attempt to answer the question, in Bill's own words, "What the ***bleep*** is W?" That is, is this moment of intent, as defined by Libet, something that can be directly perceived, experienced, measured, or reliably reported? This question thus launched our journey.

In our laboratory, there was no fancy equipment, no advanced imaging techniques. All we had were some electronic apparatuses built by our dedicated in-house electrical engineer, Al Yeck. Three bright Pomona undergraduate students, Kenton, Matt, and Jesse, along with myself, ran probably a hundred behavioral experiments and we observed something that we felt was a solid piece of evidence that would challenge Libet's findings. We pondered about it for a while, asking ourselves whether we had enough data; whether an electrophysiological component was needed to publish; could we go up against other data collected with advanced brain imaging techniques? Then, Bill came running into my office, with the Eureka! Moment on his face, excitedly announced "Let's publish this, the world needs to know!"

Our simple study, like many others from our lab, was modeled after the original Libet paradigm but with a twist (Banks & Isham, 2009). We specifically focused on the question of how Libet's W might get generated. The participants were instructed to perform a voluntary action (in our case a button push instead of a wrist flexion used by Libet). Upon the button push, they also heard a brief tone. For the ease of discussion, our variant of the Libet paradigm shall be called the Banks–Isham–Libet or BIL (a nod to Bill Banks).

The delivery of the tone was a critical addition to the study. This was important because it tested two fundamental assumptions of the Libet paradigm:

a) W represents the moment of intention to act, and
b) W causes the action

If these two assumptions hold true, introducing a tone should not change W in anyway because the intention to act necessarily precedes delivery of the tone.

What we found, however, was that our post–action tone actually had a significant influence on the W report. When the tone was presented 5 ms after the button push, W shifted from -140 to -120 ms in the direction of the tone. When the tone was presented at other time delays at 20, 40 and 60 ms, W shifted systematically toward the corresponding tone. This is a critical finding because it illustrates a violation of the assumption that W is caused by the intent to act and not other factors. As shown by our data W shifted in accordance to the time of the tone that came after an action had already taken place. This further complicated the assumption that W truly represented the moment of intention. If W were the true temporal index of intention, why then would W depend on a tone event that was caused by the supposedly intended action. If W were real, should it not be independent and free from any post-action influences?

Of course, skeptics could say that W is real but could not be accurately reported. In that case, the work like Libet's which relies significantly on this subjective report, does not provide anything informative about the temporal relationship between W and neural events related to action.

As aforementioned, we did not acquire neurophysiological data. Some may argue that without coupling this with a neurological signature, it would be difficult to compare to or challenge previous work that does involve a form of neural mechanisms. I beg to differ. Recall that we have had great contemplation about publishing our data without neurophysiological component but later decided that our data were sufficient to rescue free will from this type of experiment. That is, if the RP truly were the preparatory stage of the muscle and were independent of W, then having collected the RP data, or not, should make no difference because the RP by definition is the precursor of W. Thus, the RP should not be at all impacted by W nor by the tone that was subsequently delivered.

Although we did not work on the RP, the reader might be interested in learning more about the efforts made by other research groups to try to understand what this neural signature represents. Kornhuber and Deecke first introduced the RP as a brain activity that accompanied preparation of a motor act. Recently, however, it was shown by Trevena & Miller (2010) that the RP would be observed by simply looking at a rotating-dot clock without any plans for a motor output. This further implies that the RP may be independent of conscious urge and will. Haggard and Eimer (1999) have offered the lateralized readiness potential (LRP) as a more suitable signature related to movement preparation. Another recent work by Schlegel et al. (2013) however suggests that the LRP may be problematic for other reasons and thus may not be the optimal solution to measuring intent to act.

SUBSEQUENT STUDIES

Our first set of data was published in Psychological Science, and it was Bill's last paper as the primary author in a peer reviewed journal. Thanksgiving 2010, five months after I graduated, Bill, in his own humble way, calmly told me "I'm going into the office tomorrow and will do as much as I can to package the [remaining] data. But you are probably going to have to finish it." My first reaction, partly in denial of what I was hearing, was that he probably was just going to retire and enjoy time with his family, especially with his two little girls. But deep down, I knew this was not true. Bill would never ever retire from science if he had the choice. So, as the conversation began to settle in, I knew my journey with Bill was nearing the end. Bill passed the following April.

Sad and confused, I felt like a child who had just lost its academic parent and I had to grow up fast. The following are a series of studies that followed. Some began in Bill's lab at Pomona, some in my own program of research, but all of which were inspired by our original work in Psychological Science and fueled by the desire to honor Bill's wish to see this work through.

Representational Momentum & the Wundt Clock: Speed, Age, and Clinical Population

We continue to investigate the concerns with the Libet-type paradigm while in parallel search for a better measure for the timing of intent. One of our focuses has been on the lag in the timing of W in reference to the onset of the readiness potential. An older idea suggests that such temporal lag is in fact due to a perceptual bias while reading the clock. The Representational Momentum phenomenon (RepMo) is a motion-based bias in which the final position of a moving object is perceived as further along its expected trajectory than in actuality (Fryed & Fink, 1985). One of the key facilitators in RepMo is the speed of the moving object: the faster the motion, the stronger the RepMo phenomenon.

The original Libet paradigm, as well as in a number of succeeding paradigms like ours and the latest by Fried et al. (2011), used the Wundt rotating–spot clock in which a dot moves in circular motion along a circular clock face and completing a rotation in 2.5 seconds. The effect of RepMo on M and W reports has been investigated by Joordens et al. (2002) and by Pockett & Miller (2007). We replicated their findings, showing that speed of the moving dot could bias the reading of time W and M. In our study, we asked two groups of participants to perform the Libet paradigm. One group used the Wundt clock whose rotating spot completed one rotation in 2.5 seconds and the other group used the clock with slower moving spot, completing a rotation in 4.0 seconds. The group with the slower clock reported M to be significantly earlier (300 ms before actual time of action, compared to the classic 90 ms as observed in the faster clock and by Libet; Figure 3).

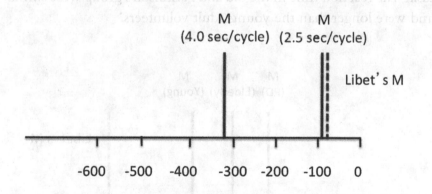

Fig. 3 Eve Isham

Averaged M reports when the clock speed was the same as Libet's clock (2.5 seconds per cycle) or when the clock speed was slower (4.0 seconds per cycle). The perceived moment of action was significantly earlier when read from the slower clock.

There are two key findings here. First, our data added to the consensus that clock speed plays a critical role in the reading of M time. Secondly, our slower clock induced an M report of -300 ms, which was now closer and approaching the classic onset time of the readiness potential (at -500 ms).

In addition to speed, it has recently been shown that older observers who have declined motion perception are less susceptible to the RepMo effect (Piotrowski and Jakobson, 2011). Persons living with Parkinson's disorder also have been reported to have declined ability in motion detection. The inefficiency in motion detection thus makes these observers less susceptible to RepMo. In the laboratory, we compared the M reports across three populations: healthy young controls, the older participants (55 and older), and those with Parkinson's disease. Note: To accommodate the ease of the task to the older populations, we used a slower-speed clock. To ensure that the participants were equated in terms of movement production ability, all participants performed a separate reaction-time task. The reaction time in the old and Parkinson's group were similar and were longer than the young adult volunteers.

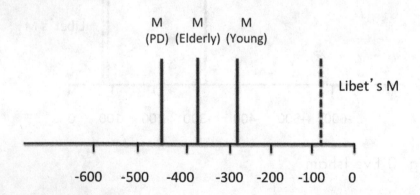

Fig. 4 Eve Isham

Averaged M reports by different populations using a slow clock (4.0sec/cycle). Older participants and those living with Parkinson's disease judged their time of action to be earlier than young adults.

Our main findings showed that the young controls, the old controls, and Parkinsonians reported M to be approximately 280, 380, and 450 ms before button press, respectively (Figure 4). Our results suggest that M (and likely W) might have been reported much later than it should be due to the RepMo bias in this timing device. This further implies that time W may also be subjected to the same effect. Given that W is commonly observed to precede M by approximately 100 ms, W could in fact be read as early as 300-500 ms when using a slower clock. This would temporally align with the onset of the readiness potential. If so, this would challenge the notion that brain activities are engaged before a conscious thought.

If the moving spot inherits a motion-based perceptual bias, it is not an appropriate timing device. Replacing the Wundt clock with a non-moving clock has been proposed as a solution. We tested this by presenting a series of 60 numbers ranging from "0" to "59" one item at a time. The number appeared at the center of the computer screen. The presentation rate was manipulated at 50, 100 and 150ms per frame. The shortest presentation rate, at 50ms, renders approximately 2.5 ms total time and was analogous to the total time of the Libet clock. By having now removed physical motion in the timing device, we predicted the M report to be earlier than Libet's M report. In contrast, we observed that the M reports were significantly *later* than Libet's M. One speculation here is that the hopeful digital clock might also have a form of mental RepMo. That is, while motion had been minimized, predictability of sequential numbers was still possible.

To minimize predictability, we randomized the numbers in the digital clock. This task proved to be difficult at the relatively fast presentation rates employed in our study that subjects failed to perform the task appropriately. A new approach is to perform a within-subjects investigation in which the observers are presented with readable displays and at the same time slowing down the Libet clock accordingly. Furthermore, the spot of light would appear at random location on the clock face instead of moving along the trajectory. This work is in progress.

How voluntary is Libet's 'voluntary' action?

If you have ever seen a magic performance, it is likely that you would see a variant of a free-choice routine. Something like "Pick a card, any card." And, the magician, who is supposedly naïve of what card has been chosen, would proceed to correctly and magically identify the chosen card. The level of success of the act is determined by how well the magician can execute and implement forcing the card to be chosen by a volunteer. The volunteer must also be naïve of the method and be convinced that he or she has the freedom of choice, uninfluenced by any other force but their own desire, to select the card. In such a case, despite the belief of the volunteer, free choice is illusory.

What about the freedom to act in the Libet's paradigm? Did the participants feel they have the choice? Did they feel that their own intention drives each wrist flexion? Although Libet told them that they did not need to do anything if they did not feel like it, Flanagan (1996) suggests that the kind of action Libet asked the participants to perform would be better classified as automatic than willed because the participants executed the action in compliance with the experimental instruction. Drawing parallelism to magic, it seems that the experimenter would already have asked the subjects to act a certain way, and not completely at will.

In addition to Flanagan's concern, we noticed something in the Libet paradigm that seems to further limit the exercise of free will. At the time of this writing, I have consulted with several scholars about the answer to my question. So far, no one knows. The big question is: why did the subjects in the Libet paradigm have to wait until the second clock rotation to act? Did the experimenters know something about the difference between the second and first rotation clock readings? Critically, if subjects had felt the urge to act during the first clock rotation, but had to wait, how voluntarily was this action?

We conducted a simple study to try to answer these questions. In our study, we asked the participants to perform a voluntary keypress while keeping time using the Wundt clock. Importantly, we omitted

Libet's instruction about withholding action until the second rotation. Instead, we encouraged the participants to press the button at any time while the clock was running. Our preliminary data showed that that majority of the participants (10 out of 14 participants total) had a higher tendency to press during the first clock rotation than the second. It seems that when given the opportunity, the participants did not wait.

A more critical question though is whether the restriction to act in the second clock rotation posed any concerns to the validity and the results in Libet-type studies. Thus, we compared the reported time of M when the action occurred during the first and the second clock rotations. We observed that the perceived moment of action (Libet's M) was reportedly earlier when the keypress occurred in the first rotation compared to the second rotation. The difference was 60 ms. While some might feel this temporal difference of 60 ms negligible, it is also important to keep in mind that a number of mental and computational processes are completed sooner than that. For instance, in the context of Libet, this would be sufficient to veto an action.

This leads to the next question of whether the 60 ms is functionally valuable or whether it is a result of yet another motion perception bias. Some might say our visual system requires some time to adjust or to adapt to visually track the moving dot on the clock face. This subsequently might have resulted in the negative lag observed in the first clock rotation when visual adaptation has just been initiated. If so, then one could further argue that the temporal reading would be more accurate after having control of the tracking of the moving dot. On the other hand, one could also propose that this is another variant of the RepMo phenomenon. When the trajectory of motion is unfamiliar or to-be-determined, such as in the first rotation, it seemed that the participants might be less susceptible to overestimate the position of the moving dot.

On the assumption that the observed difference is of functional significance and free of perceptual influences, then the validity of the Libet-type paradigm is once again a subject of criticism: Was it appropriate to place a restriction on when the action could be executed when the experiment was designed to address the actions of *voluntary* nature?

Intentional Binding

Another line of investigation that illustrates that the perceived time of action could be an index of consciousness is on the phenomenon called "intentional binding." The term was originally coined by Haggard et al. (2002). They proposed that the perceived duration between an action and its outcome varied with the degree of perceived intentionality. If the action is intended, the duration is judged to be shorter, hence the term intentional binding. In contrast, a longer perceived duration is reported for an unintended action (i.e., when the action was incidental or when induced by exogenous stimulation such as the transcranial magnetic stimulation procedure).

Haggard and colleagues' work is innovative and in the same manner as Libet's, has established yet another connection between science and the philosophical perspective on intentionality and agency. And as in the same manner as Libet's, it too received wide support as well as skepticism. Moore et al. (2010) supported intentional binding by extending it into the clinical realm, proposing that the lack of intentional binding effect in Parkinson's disease is indicative of minimized lack of agency in this clinical population. On the other hand, Eagleman and Holcombe (2002) proposed that this temporal binding was not driven by intentionality but by causality. That is, when two events are perceived as causal, but not necessarily intentional, these causal events are also perceived to be closer in time. Buehner and Humphreys provided evidence that supported this claim (2009, 2010).

Moreover, we too have found that intentionality may not be the true basis of this temporal attraction (Isham et al., 2011a). In our study, participants were asked to race against a confederate at pushing the button in response to a random flash. The naïve participants were told that whoever pressed their button first would elicit their own tone, pre-assigned before the experiment. Unbeknownst to the participants, the confederate's button was a decoy and did not generate a tone. The tonal effect was generated directly from the participant's button push. The tone was randomly presented by the computer. If intentionality were the true basis of the temporal binding, we would expect the M report to be later and closer to one's own tone (i.e. intended effect) than the M report of the trials in which the confederate's tone (i.e., unintended effect) is heard.

This was not the case. We observed the opposite of what would have been predicted by the intentional binding. Instead, the M report was earlier for trials in which the participant's own tone was heard than for trials in which the confederate's tone was heard, suggesting a temporal dilation rather than a temporal compression in the intended trials. This thus implicates that intentional binding may not be the best index for intention and agency.

We extended the study further and tested whether the observed effect was influenced by external factors such as the presence of another player. Joy Geng, my postdoctoral mentor, and I, thus conducted a follow-up study to test whether competing against oneself would still facilitate a temporal dilation seen in the Isham et al. study. The results are consistent with the previous finding, suggesting that the timing of an intended action does not necessarily bind to the timing of an intended outcome (Isham & Geng, 2011). That is, the perceived time of action is not driven by intentionality, nor is it directly perceived for that matter. Instead, time telling is done retrospectively, and after having evaluated related and relevant information.

Confidence ratings

The BIL paradigm with inserted delayed tones also launched a set of questions itself. One of these was on how susceptible people are of the deceptive delayed tones. Were the observers completely deceived that there was no delay, or whether they were somewhat aware but could not actively avoid the binding of the auditory tone with the other attributes of the event? Work by Rigoni et al. 2011 provided evidence that while their observers behaviorally appeared to be fooled by the tone (i.e. producing the same effect as what we reported) although their brain activities (i.e. error detection negativity in ERP) indicated a stronger signal discrepancy when there was a greater temporal difference between the time of action and the time of the delayed tone.

So while Rigoni et al. showed that the *brain* could detect some sort of discrepancy, and some behavioral output appeared to suggest the opposite, what about the mind? If we were to assess for the mind's response to the action–tone discrepancy, would the data reflect that of the brain or of the behavior, or perhaps something in between? To answer this question, my student, Tiffany Wall, and I asked our research participants to perform the BIL task and to report M and W. In addition, they also rated how confident they were of the accuracy of their reports. Confidence ratings have been applied to assess for the subjective perspective of one's cognitive performance including list learning, source memory, and others (e.g., Yonelinas, 1994). Here we asked whether confidence ratings would serve as indicator of consciousness (Dienes & Scott 2005).

Our results (Isham & Wall, under review) suggest that behaviorally participants relied on the post-action tone to report Libet's moments of W and M. This is consistent with previous observations reported by us (Banks & Isham, 2009, 2010) and by Rigoni et al. (2011). However, when the observers were asked to assess their sense of accuracy of their own M or W reports, they felt that they were more accurate when the tone was given at a shorter delay as opposed to a longer delay. These results suggest that the observers were temporally aware of the timing of their action, but were unable to explicitly separate the action event from the tone event in the W and M reports. This adds to the evidence raised in this essay that we are not able to accurately report when we are actually conscious of our own action.

At this point, you might wonder what kind of information is influential to time telling. It seems that temporal reports are retrospectively processed based on both relevant environmental factors and internal time keeping. While work is underway to further assess this temporal coupling of action and its surrounding events, it also presents a critical concern when studies rely on accurate time perception as an index of consciousness. You might wonder if there might be a better way. Based on our research thus far, it seems it would be best to avoid using subjective reports to index moments of consciousness. Below is one possible solution I can offer, but I am sure there are other ways that one could think of.

When there's a will, there's a way: Timing of Eye Movements and "Free" Choice

One of the goals of my research is to find a more rigorous method to assess for the timing of intent. If this moment of W were real, then it must manifest itself in some form. Thus we turned to a subjective decision-making paradigm and examined the changes in eye movement characteristics as a function of choice (Isham & Geng, 2013). Specifically in our study, the participants were asked to view two novel black and white patterns while their eye movements were being monitored. They indicated the more aesthetically preferred item by pressing a keyboard button that corresponded to the choice. Subsequently, we performed a time series analysis and compared the accumulating time spent fixating on the two options before the manual response. We found that people spent more time looking at the chosen item than at the unchosen item. Critically, this divergence occurred as early as 2500–3000 ms before the manual response. We propose that this moment of divergence is potentially related to W.

Speculatively speaking, if the moment of divergence in looking time closely represents the moment of urge or desire, then it would further raise two critical concerns. One, the onset of this moment would be significantly earlier than what was observed by Libet's method, which he termed W. Moreover, the moment of divergence hypothetically would precede the onset of electrophysiological components known to be involved in motor preparation (such as the readiness potential and the lateralized readiness potential). Employing the same logic as the one used in the argument against free will, this hypothetical result would in fact suggest that the time of urge causes the brain to engage in a motor act, and not the other way around. Of course, verification is needed.

In addition, this speculation about W also leads to the perspective that perhaps W is not a discrete event. Instead the conscious moment of urge gradually develops or builds up. Our research, as well other work in the drift diffusion or accumulator model would be inline with this perspective. This view would again challenge the perspective that the moment of urge is an isolated, discrete event.

THE END (FOR NOW)

It has been more than three decades since Libet and colleagues conducted the seminal study that, to some, has threatened free will. I have humbly presented in this writing a small collection of studies which question the validity of this type of experiment. I sincerely hope that this essay has helped the reader become more informed about this line of research, its findings, and the subsequent interpretations. I must note that our own studies are not without flaws. And, I would encourage anyone to take our studies further should the new potential results help advance science.

In closing, I must also emphasize that while we have illustrated some of the concerns with the Libet paradigm, I nevertheless respect the pioneer work by Libet and colleagues. For it truly advanced scientific thinking and engaged scholars to think more interdisciplinarily. Libet's work also gave birth to new lines of studies such as neuroethics and neurolaw. And of personal importance, it created an opportunity for Bill Banks and I to partake a research journey together.

REFERENCES

Banks, W.P. & Isham, E.A. (2009). We infer rather than perceive the moment we decided to act. Psychological Science, 20, 17-21.

Banks, W. P., & Isham, E. A. (2011). Do we really know what we are doing? Implications of reported time of decision for theories of volition. In W. Sinnott-Armstrong & L. Nadel (Eds.), *Conscious Will and Responsibility* (pp. 96–125). New York: Oxford University Press.

Banks, W.P. & Pockett, S. (2007). Benjamin Libet's work on the neuroscience of free will. In M. Velmans & S. Schneider (eds.), *The Blackwell Companion to Consciousness*. Blackwell.

Buehner, M. J. (2010). Temporal binding. In: Nobre, A. and Coull, J. eds. *Attention and time*. Oxford: Oxford University Press, pp. 201-211.

Buehner, M. J. and Humphreys, G. R. (2009). Causal binding of actions to their effects. *Psychological Science*, 20, 1221-1228.

Eagleman, D.M. & Holcombe, A.O. (2002). Causality and the perception of time. *Trends in Cognitive Science*, 6(8), 323-325.

Dienes, Z., & Scott, R.B.(2005). Measuring unconscious knowledge: distinguish structural knowledge and judgment knowledge. Psychological Research, 69, 338-351.

Flanagan, O., (1996). Neuroscience, agency and the meaning of life. In *Self-Expressions*. Oxford: Oxford University Press.

Fried, I., Mukamel, R., & Kreiman, G. (2011). Internally generated preactivation of single neurons in human medial frontal cortex predicts volition. *Neuron, 69,* 548-562.

Freyd, J.J., & Finke, R.A. (1984). Representational momentum. *Journal of Experimental Psychology: Learning, Memory and Cognition, 10,* 126-132.

Haggard, P., Clark, S., & Kalogeras, J. (2002). Voluntary action and conscious awareness. *Nature Neuroscience, 5,* 382-385.

Haggard, P. & Eimer, M. (1999). On the relation between brain potentials and the awareness of voluntary movements. *Experimental Brain Research, 126,* 128-133.

Isham, E.A., Banks, W.P., Ekstrom, A.D., & Stern, J.A. (2011). Deceived and distorted: Game outcome retrospectively determines the reported time of action. *Journal of Experimental Psychology: Human Perception and Performance, 37,* 1458-1469.

Isham, E.A., & Geng, J.J. (2011). Rewarding performance feedback alters reported time of action. *Consciousness and Cognition, 20,* 1577-1585.

Isham, E.A., & Geng, J.J. (2013). Looking time predicts choice but not aesthetic value. *PLoS ONE. DOI: 10.1371/journal.pone.0071698*

Joordens, S., Spalek, T.M., Razmy, S., & van Dujin, M. (2004). A clockwork orange: Compensation opposing momentum in memory for location. *Memory & Cognition, 32,* 39-50.

Kornhuber HH, Deecke L (1965) Hirnpotentialänderungen bei Willkürbewegungen und passiven Bewegungen des Menschen: bereitschaftspotential und reafferente Potentiale. Pflügers Archiv. *European Journal of Physiology, 284,* 1–17.

Libet, B. (2011). Do we have free will? In W. Sinnott-Armstrong & L. Nadel (Eds.), *Conscious Will and Responsibility* (pp.1-10). New York: Oxford University Press.

Libet, B., Gleason, C. A., Wright, E. W., & Pearl, D. K. (1983). Time of conscious intention to act in relation to onset of cerebral activity (readiness-potential). Brain, 106, 623–642.

Libet, B., Wright, E.W., & Gleason, C.A. (1983). Preparation-or intention-to-act, in relation to pre-event potentials recorded at the vertex. *Electroencephalography and Clinical Neurophysiology, 56,* 367-372.

Moore, J. W, Schneider, S. A., Schwingenschuh, P., Moretto, G., Bhatia, K. P, & Haggard, P. (2010). Dopaminergic medication boosts action-effect binding in Parkinson's disease. Neuropsychologia, 48(4), 1125–1132.

Piotrowski, A.S., & Jakobson, L.S. (2011). Representational momentum in older adults. *Brain Cognition, 77,* 106-112.

Pockett S and Miller A (2007) The rotating spot method of timing subjective events. Consciousness and Cognition 16, 241-254.

Rigoni, D., Kuhn, S., Sartori, G., & Brass, M. (2011). Inducing disbelief in free will alters brain correlates of preconscious motor preparation: The brain minds whether we believe in free will or not. *Psychological Science, 22,* 613-618.

Schlegel, A., Alexander, P., Sinnott-Armstrong, W., Roskies, A., Tse, P.U., & Wheatley, T. (2013). Barking up the wrong free: readiness potentials reflect processes independent of conscious will. *Experimental Brain Research, 229,* 329-335.

Soon, C. S., Brass, M., Heinze, H., & Haynes, J. (2008). Unconscious determinants of free decisions in the human brain. *Nature Neuroscience, 11,* 543-545.

Trevena, J. & Miller, J. (2010). Brain preparation before a voluntary action:

evidence against unconscious movement initiation. *Consciousness and Cognition, 19,* 447–456

Yonelinas, A.P. (1994). Receiver-operating characteristics in recognition memory:

Evidence for a dual-process model. Journal of Experimental Psychology: Learning, Memory and Cognition, 20, 1341-1954.

ACKNOWLEDGEMENTS

All the projects discussed in this writing would not have been possible without the help of so many individuals. I would like to first thank all of those who were significantly involved in these projects (Tiffany Wall, Joseph Butler, Rachael Gwinn, Farhan Sareshwala, Al Yeck, Kenton Hokanson, Jesse Huston, Matthew Maciallo, Jessica Stern, Krystal Wulf, Joy Geng, and Arne Ekstrom) as well as those who have provided guidance along the way (Andrew Yonelinas, Steve Luck, Jeff Miller, Tim Hubbard, Elizabeth Disbrow, Dale Burger, Cathy Reed, Kathy Pezdek). A special thank you to my editor, Ingrid Fredriksson, who never gave up on this writing. And Bill Banks, thank you for *everything!*

THE EXTENDED MIND

Rupert Sheldrake

I AM SPEAKING NOW ON the theme of the extended mind, by which I mean the mind beyond the brain. The question I'm addressing is 'Where is the mind?' and for most people in the academic world it seems pointless asking that question, because they think they already know the answer. The mind is in the brain. Mental activity is brain activity. This is the central assumption of the materialist philosophy or the materialist world view. It has not been proved that the mind is nothing but the brain, but is assumed. And it's assumed on the basis that changes in the brain can affect the mind. Damage someone's brain, and it affects their mental activity. Give them a lot of alcohol to drink, they behave differently. These things have been known for centuries if not millennia. But they don't prove that the mind is nothing but the brain. They prove that the brain has something to do with mental activity. And what I'm going to suggest this morning is that minds are much more extensive than brains, they stretch out far beyond brains – through fields.

Now, I will leave the question of what kind of field till later. I am not talking about regular gravitational or electromagnetic or quantum fields, but about another kind of field. But let me first say that the idea of fields extending beyond material bodies, is now completely mainstream in science. The field concept was first introduced by Michael Faraday in the 1840s, which the idea of the electric and magnetic fields. The magnetic field is inside a magnet, the material object, but stretches out invisibly beyond it. It's a region of space, the most general definition of a field is a region of spatial influence, a region of influence in space. The gravitational field of the Earth is inside the Earth, but stretches out invisibly beyond it, holding the moon in its orbit and holding us down on the floor, otherwise we'd be floating around. We can't see it, but it's a region of influence in space. The electromagnetic field of your mobile telephone is inside the mobile telephone and stretches out invisibly beyond it. We are surrounded electromagnetic transmissions from radio and mobile phones, and what I'm suggesting is that our mind are normally rooted inside our brains, but extend far beyond them through mental fields.

Now the debate about that the nature of the mind has been coloured for more than 300 years by Cartesian dualism. René Descartes, as everyone knows, made a radical separation of spirit and matter, mind and body. The mind, Descartes thought, was not in space and time. Only humans had minds in the whole natural order, animals didn't and everything else in nature was mechanical. The only other beings with minds, or spirits, were God and angels. So God, angels and human minds were not in space and time. But bodies, or matter, the whole of material reality, were in space and time. They were *res extensa*, extended things.

The materialists in the 19th century didn't like Cartesian dualism, so they denied the spirit pole and said the only reality is matter. Materialism is a philosophy that arises through the denial of the spirit pole of dualism. It is a very unsatisfactory dualism, but materialism is even more unsatisfactory, in my view. What I'm suggesting is a third view, which is neither Cartesian view of disembodied spirit outside space and time, or materialism that's just matter. This third view says the mind is extended in space, it's extended all around us, it's extended throughout our brains and bodies. It's extended through fields. The same applies to animal minds as humans minds. This is not a special hypothesis of "human spirit". It's a theory of the nature of mental activity extended in space.

Now the easiest way to see what I'm talking about is to think of the nature of vision. What's happening when you see something? Like what's happening when you see me now? Well, you all know the standard textbook answer: light is reflected from me, travels through the electromagnetic field, it enters your eyes, inverted images form on your retinas, changes occur in the cone cells, impulses travel up the optic nerve and changes occur in various regions of the brain, which could be described in great detail by fMRI and other scanning techniques. We know more about the changes in the brain than ever before. That's very good as far as it goes. But what it does is to just describe the changes that happen in the eyes, the optic nerves and brain when you see something. It doesn't account for two essential features of the vision. First of all that you are conscious of what you see. This is an example of the hard problem. You ought not to be conscious of anything, if matter is unconscious and you are made of nothing but matter. And many philosophers of mind have been trying to pretend that we're not conscious, it's just an illusion.

But the question I want to talk about is not 'the hard problem', but where these images are. When you see an image of me, where is your image of me? Where is your image of the rest of this room and of everything you are seeing?

The standard view, the materialist view, is that it's all in the brain. All your experiences are in the brain. Somehow, when you see things changes occur in the brain, electrochemical changes occur in nerves cells, and electromagnetic changes occur over large regions of the brain.

And then somehow, your mind or brain produces a 3D virtual reality display inside your head, which contains everything you're seeing in three dimensions in full colour. It's all supposed to be inside your head. How the brain, or the electric fields in the brain generate this 3D virtual reality display, is left entirely unexplained. And no one has ever seen a 3D virtual reality display in a brain. When people open up brains for surgical operations, they don't see full colour 3D displays of what that person is experiencing. But this is the official view. And according to that view, somewhere inside your brain there's a little Rupert standing here in full colour, in three dimensions, everything else you are seeing is inside your brain. When you look at the sky, the sky you are seeing is inside your brain. A recent paper in "Brain and Behavioural Sciences" was entitled "Is your skull beyond the sky?". The author, who is a materialist philosopher of mind, answered the correct materialist answer. Yes, your skull is beyond the sky that you are seeing, because all your experiences are inside your head.

This materialist theory view is taken for granted by materialist philosophers and by most other people because that's what they've been taught to believe. But it does not correspond with our experience. We experience the images we see as being outside ourselves. I experience what I'm seeing of you not as being inside my head. Everything I see seems to be out there. And what I'm proposing is a hypothesis so simple that it's hard to understand. Your image of me is located exactly where it seems to be, where I am standing. It's outside your head. It's in your mind, but not in your brain. You project out images of everything you're experiencing. Light comes in, changes occur in the brain and images are projected out to where they seem to be. Sometimes they're projected out where things are not and then you have an illusion or hallucination. That can happen. But usually, they correspond to where things are and if they didn't we wouldn't have got here this morning. We would've been bumping into things and we'd be completely dysfunctional.

I claim no originality for this hypothesis. It's the theory of vision put forward by Plato, it's the theory of vision put forward by Euclid, the great geometer, who was the first person to explain mirror images. Euclid argued that the light is bent, that the light reflected of the mirror, comes into your eyes, but you project your virtual images that you are seeing and because they are virtual, they go straight through the glass and you see them behind the mirror. So every time you look in a mirror, you are seeing your own visual projections. This is also the view taken by traditional people all over the world and it's the view that children have in our own culture. Jean Piaget, the developmental psychologist, showed that until the age of about

10, children think that vision involves an outward projection of images. That's why in Superman comics, you see rays go out of his eyes. And Roald Dahl's story Matilda appeals to children very strongly, because she has eye beams that can move things. But as Piaget said, by the age of ten or eleven the child learns the "correct" view that images are invisible things located inside the head. So we have all been brought up to deny our most immediate experience of the world in favour of an untested hypothesis, that's nothing but an assumption that dates back to the 17th century.

So what I'm suggesting is that this is how vision works. Our minds stretch out from our brains in the act of vision. This is now a fairly fashionable position within philosophy of mind. It's sometimes called radical externalism and this is now a hot topic of debate. Until fairly recently, most people took for granted that it's all inside the head, but this is no longer the case. There a much wider debate about this. But it's usually treated as a philosophical debate. I treat it as an empirical hypothesis. If our minds stretch out beyond our brains and if there are fields, however we conceive them, mental fields or perceptual fields that stretch out containing these images.

When we look at something, our mind may affect what we look at, because it reaches out in a sense to touch it. If I look at a distant star, my mind may reach out over astronomical distances and probably years backwards in time as well. If I look at you from behind, and you don't know I'm there, the question is could you pick up the feeling that you are being looked at by me. Just because my mind is reaching out to touch you, even if I'm looking through a window, even if there's no sound and no smell. Is it conceivable that people can feel when they are being looked at from behind?

This would be an empirical test. Now as soon as you ask that question, you realise this is an everyday phenomenon. It's not paranormal or weird, it's not strange and bizarre. It's an everyday phenomenon experienced by most people. Surveys have shown that over 90 % of the population have had this experience, including children. There is a slight sex difference; more women than men have experienced being stared at. More men than women have experienced staring at others and making them turn around. But nevertheless, almost everybody has had this experience. I call it the sense of being stared at. The scientific name for it is scopaesthesia. *Scop* as in looking, *aesthesia* as in feeling.

So does it really happen? Everyone, who has been to university and trained in the sceptic materialist world view, has a ready answer. "No, of course it doesn't really happen. You just turn around all the time and if someone catches your eye, you imagine that they've made you turn around, but you just forget the millions of times you are wrong."

This is the standard armchair sceptical explanation. It has inhibited research on the subject for a long time. This is such a taboo topic that until about 1985, there were only been four papers in the published literature on this phenomenon. All four had shown positive effects, but two were by sceptics who tried to explain them away. Since the 1980s there has now been quite a lot of research on this subject. Many of the experiments are so simple that a child can do them and indeed many of them had been done by children. They'd been done extensively in schools in Britain, Germany and in the United States.

There have now been hundreds of thousands of these trials. The evidence is summarised in the review paper of mine in the *Journal of Consciousness Studies* in 2008 (Vol12, pp10-31). You can find the full text on my website, so I'm not going to go into the experimental details. Basically the experiment involves the subject who wears a blindfold like an airline blindfold to eliminate peripheral vision, the looker sits behind them and in a randomized series of trials either looks or doesn't look. The beginning of the trial is signalled by a beep or a click. The subjects, within 10 seconds, have to guess whether they are being looked at or not. They're right or they're wrong. By chance they would be right 50% of the time. In this huge series of experiments, the average hit rate is about 55%. Not a big effect. But massively significant statistically. Some people, of course, are more sensitive to others. But this is an overall effect of unselected subjects. These experiments have been done all over the world, including here in Sweden. Some people in this room took part in the experiment I ran about 15 years ago.

Some of these experiments have been done through one-way mirrors and through windows to eliminate any lingering possibility there could be sounds or smells giving clues. And this experiment, a computerised version, has been running in the Amsterdam Science Museum, the NEMO centre, for the last 20 years. One of the biggest experiment ever conducted. With overwhelmingly positive statistically significant results. The most sensitive subjects in that experiment, and indeed in my own, are children under the age of 10. Most adults get rather thick-skinned. You know, we're exposed to lots of people in modern cities. Children are more sensitive and they are very interested in the phenomenon, which is why it's been quite easy to do it in primary schools. School teachers are also interested because one of the tricks of their trade is to stare hard at a naughty boy, who has turned around and very quickly the children turn back again.

These experiments have also been done using close circuit television. People in another room, even in another building, are looked at or not looked at in a randomised sequence. And there are physiological changes in the skin conductance, showing that they're physiologically aroused when they're being stared at by someone in a different room.

A few weeks ago, I launched, for the first time, an online staring experiment that I invite all of you to try if you can. It involves two people, you can be in distant places on the computer. It uses your computer camera in a SKYPE-type photography. If you're the subject you're looked at by another person, or by many other people in different parts of the world. In a series of trials you will have to guess whether you're being looked at or not. I don't know the results yet, because the experiment has only been running three weeks so far, but do have a go. It's on my website www.sheldrake.org

It turns out that animals also are sensitive to being stared at. Many wildlife photographers and hunters have noticed this. And I think that this phenomenon is easily explained in evolutionary terms in the context of predators and prey. Any prey animal that could feel when a predator was looking at it would have an advantage compared a the prey animal that could not tell when a predator was looking at it. It's a useful ability. It is also well-known in the security industry. I've interviewed a lot of people, who spend their time looking at other people; surveillance officers, store detectives, the Heathrow drug squad and so forth. Among such people it's completely taken for granted that this happens. Even with CCTV. In the martial arts, they have ways of training people to become more sensitive to being looked at from behind. It's useful to know if someone's looking at you from behind and creeping up to attack you. It's probably widespread in the animal kingdom.

Nevertheless this phenomenon hasn't been investigated by naturalists or by animal behaviour people, because the taboo on this subject is so immense, because the conventional scientific way of thinking is hemmed in by a whole series of dogmas and beliefs, which make people feel that it is dangerous for their careers to study these subjects.

Similar taboos apply to telepathy. Telepathy literally means 'distant feeling'. *Tele* means distant and *pathy* means feeling, as in sympathy or empathy. And I think telepathy is also basically a biological phenomenon that animals have, not just us humans, but many species have this. And I think telepathy is a function of the bonding that occurs between animals in social groups. Social animals by definition live in societies and have to interact with other members of their social group. I think these groups are coordinated by social fields, which are a kind of morphic field. I am not talking about the nature of these fields today, just mentioning them. This is a basic holistic model of how nature is organised. Everything in nature is organised in a nested hierarchy, like subatomic particles in atoms, atoms in molecules, molecules in crystals. At each level the whole is more than the sum of the parts. The wholeness at each level, that which organises the wholeness is what I call a morphic field. Morphic from the Greek word *morphe* meaning shape or form.

Social animals are within groups, within the field of the social group. So I think social groups of animals are indeed organised by social fields, which are morphic fields, invisible organising fields analogous to magnetic fields, which are also invisible organising fields. What I'm suggesting is that animal groups have these fields that link together the members of the group and that the bonds between the members of the group link them through these morphic fields of the group, and can act as a channel of communication. If members of the group separate from other animals then the field doesn't disappear, it stretches. For example, if wolves have young, have cubs, they leave the cubs in the den and then the adult wolves go out hunting, ranging over hundreds of kilometres in the case of northern Canada. They go enormous distances. The bond between them is not broken, but it stretches. And observations by naturalists of wolves suggest that they can respond to what is happening to other members of their group many miles away, far beyond the range of hearing or smell. I think this is because there's a kind of telepathic connection between them. I think telepathy is a normal form of communication within animal groups. They don't have mobile phones. They don't have the internet. They don't have text messages. And in order to coordinate the activity of separated members of the group, telepathy is what they use. It's a useful ability, and that is why it has evolved. That's my hypothesis.

It's analogous to quantum entanglement. I don't say it's the same as quantum entanglement, but it's worth mentioning the analogy. If two particles have been a part of the same quantum system, and they move apart, they remain entangled or non-locally connected or non-separable according to the principles of the quantum mechanics. A change in one leads to an instantaneous change in the other. And this was opposed by Einstein, who called it "spooky action at a distance". But experiments have shown that spooky action at a distance really happens. Quantum theory was right, Einstein was wrong. This does not fall off with distance, that's one of the interesting things about quantum non-locality. Unlike say the intensity of gravitation, which falls off according to an inverse square, quantum non-locality does not. Telepathy is more analogous to quantum non-locality than to gravitation or the diffusion of light from a point source.

What I want to do now is to talk about experimental investigations. Because I think it's common in animals, I started by looking at animal telepathy. I think that telepathy, to summarise this point, is normal, not paranormal. Natural, not supernatural. I think it's an animal ability that we share with many animal species. So because I think it's a natural part of animal life, I started my investigations by looking at the animal we know best; dogs, cats, parrots, horses — domestic animals that form bonds with their human owners. I started as a biologist with natural history, so I collected stories from people about their experiences. I now have more than 5,000 in an animal database. These stories provide a natural history of people believe about their animals. They don't prove these beliefs are true, but they make it clear that certain phenomena occur over and over again.

For example, in this database, I have more than 200 cases where people say that their cats know when they're planning to take them to the vet. The cats disappear. Cats hate going to vets on the whole, and many people have the problem that the cat simply vanishes when they want to take it to the vet. So after this has happened a few times, people try to avoid letting the cat know when they're going to the vet. They don't let it see the cat basket, they try to avoid mentioning the word vet, but the cat still seems to know and some people in despair, ring the vet from work to make an appointment so the cat can't overhear the conversation. Then they swing by home after making the appointment to pick up the cat to take it to the vet and it's just not there. It's hiding. We have heard so many of these stories, we did a survey of veterinary clinics in north London, from the north London yellow pages, 65 veterinary clinics. We rang them all up and asked "do you ever have a problem with people missing appointments with cats?". 64 out of 65 said "yes, it happens all the time." and the 65th one said "No, it happened so often, we have given up the appointment system for cats.".

The easiest to test of the many abilities of cats and dogs that suggest telepathic abilities is dogs and cats that anticipate the arrival of their owners. Many people have had the experience of their dog or cat seeming to know when the owner is coming home. They go and wait at the door or a window. I discussed this with my sceptical friends, and one of my close friends in England is an outright sceptic, Nicholas Humphrey. I always discuss my experiments with him first, because I prefer to get the sceptic input at the beginning rather than at the end. So when I said to him that I was interested in dogs that know when their owner is coming home, and instead of saying that this doesn't happen, he said "Oh, my dog used to do that. My mother always knew when I was coming home about half an hour in advance". So I said to him "But Nick, your dog couldn't possibly have heard your car from twenty miles away, the other side of Cambridge with the motorway in between. ". He said "Oh, on the contrary. It just shows what sharp hearing they've got". So together we thought up a simple experiment, which I've now done many times.

We film the place the dog waits, we have people go at least 8 km from their home. They come home at randomly chosen times, that they don't know in advance. We pick the time at random and ring them up on a mobile phone to tell them when to come. And to avoid familiar car sounds, they come in unfamiliar cars, usually taxis. The most expensive aspect of this research are the taxi fares. We also did surveys to find out how common this behaviour is. We found that about 50% of dogs know when their owners are coming home.. Fewer cats do it, about 30% of cats. I don't think cats are necessarily less sensitive, some of them are just less interested. Including, unfortunately, my own.

In our experiments, we've done many of these videotaped experiments and again all of this is in peer-reviewed journals, you can read the texts on my website. There are dozens of papers on these kinds of things I'm talking about.

The same principles of telepathy I think apply to humans. Again with humans, I started not from laboratory parapsychology, but from a collection of stories, and now have about 5,000 stories from humans about their psychic experiences. One group of stories were from mothers and babies. Many mothers claimed they know when their baby needs them. If they're nursing mothers, their milk lets down just before... when they're away from the baby, if they feel their milk lets down, their breasts squeeze out milk, it's an oxytocin mediated milk let-down response. Most mothers assume their baby needs them and they ring home on their mobile phone. And they're usually right. I have had hundreds of stories and I did surveys of nursing mothers with the help of a midwife, who worked for me, showing this is a very common phenomenon.

It's a very useful thing for a mother to know when her baby needs her at a distance. Before the invention of telephones, mothers, who knew when their baby needed them, would have had babies that survived better than mothers that didn't know when their baby needed them. Before the invention of telephones, telepathy was the only way of alerting people to need at a distance.

I've done a series of empirical studies with mothers and nursing babies, I've summarised these in my book "The sense of being started at and other aspects of the extended mind". I summarised the animal research and many other animal experiments in my book "Dogs that know when their owners are coming home and other unexplained powers of animals".

By far the commonest kind of telepathy in the modern world occurs in connection with telephone calls. Surveys show that over 80% of people have had telephone telepathy experiences. You think of someone, they then ring and you say "that's funny, I was just thinking of you". This sort of thing has been going on ever since the invention of the telephone, but for a hundred years armchair sceptics managed to inhibit research by the standard armchair sceptical argument "Oh yes, well, you may think it's telepathy, but in fact what's happening it's just a coincidence. You think about people all the time, if one of them rings then you think it's telepathy and you just forget the millions of times you're wrong.". You've all heard that argument, I'm sure. Many of you may have made it. I used to myself, when I was anti-telepathy sceptic during my atheist phase in my twenties. But then I asked these sceptics "where's the data". You know, in science it's not enough to have a hypothesis, you need the data. I found no data at all, no one had ever investigated it.

I've set up a series of experiments, which are quite simple. They're statistical experiments. We find people, who say this happens to them quite often. They sit at home being filmed with a landline telephone, with no caller ID display. They give us the names of four people they know well, we pick one of them by the throw of a die or by a random number generator, ring them up and ask them to phone their friend. The phone rings and before they pick it up, they have to guess who it is. "I think it's Mary", they pick it up "Hello, Mary". They're right or wrong. By chance they'd be right 25% of the time, one in four.

Well, in these experiment, the actual hit rate in the filmed trials was 45%. Compared with 25% by chance. This phenomenon is more significant than the suggestion of the Higgs boson.

In these kinds of experiments we've compared familiar and unfamiliar callers. Telepathy occurs between people, who are emotionally bonded, not with strangers. One of the problems with a lot of the parapsychology tests is that they take complete strangers. In real life it happens between husbands and wives, mothers and babies, parents and children, best friends, close colleagues etc.

In these four telephone tests, four callers, we had two familiar and two unfamiliar callers in a series of trials and with the familiar callers. With the familiar callers the hit rate was over 50%, with the unfamiliar callers it was only just above chance, which is what you'd expect given the nature of telepathy. That, by the way, is why when John Joe McFadden said yesterday "Why don't poker players know what the other person's thinking?", the answer is that they are not usually bonded. They're usually highly competitive. In bridge, where people play in teams, there's a lot of evidence for telepathic communication between partners in games of bridge. So it happens in bridge, but not in poker.

And we've done experiments with email telepathy with very similar results to telephone telepathy. Way above chance, the scores. Highly significant. We've done the same thing with SMS messages and it's exactly the same kind of results and principles. Again, you can find all these papers on my website.

So in summary, I think there's a lot of evidence that our minds are extended beyond our brains. Through attention and intention.

REFERENCES

Sheldrake R. 2004. THE SENSE OF BEING STARED AT (cited 2017 Jun 25) Available from: https://www.amazon.co.uk/Sense-Being-Stared-At-Sheldrake/dp/B00C7GIASY/ref=sr_1_1?s=books&ie=UTF8&qid=1495122702&sr=1-1&keywords=THE+SENSE+OF+BEING+STARED+AT

_____, 2011. DOGS THAT KNOW WHEN THEIR OWNERS ARE COMING HOME (cited 2017 Jun 25) Available from: https://www.amazon.co.uk/Dogs-That-Their-Owners Coming/dp/B00FKYXWRE/ref=sr_1_2?s=books&ie=UTF8&qid=1495122764&sr=1 2&keywords=dogs+that+know+when+their+owners+are+coming+home.

_____, 2012. THE PRESENCE OF THE PAST (cited 2017 Jun 25) Available from: https://www.amazon.co.uk/Presence-Past-Morphic-Resonance-Memory-x/dp/1594774617/ref=sr_1_2?s=books&ie=UTF8&qid=1495122826&sr=1-2&keywords=the+presence+of+the+past

_____, Journal of Consciousness Studies, 2005. (cited 2017 Jun 25) Available from: http://www.sheldrake.org/files/pdfs/papers/JCSpaper1.pdf

_____, Smart P. Journal of Parapsychology, (cited 2017 Jun 25) Available from: http://www.sheldrake.org/research/telepathy/videotaped-experiments-on-telephone-telepathy.

MILLENNIAL SCIENCE

The Imminent Age of Discovery's 'Conscious' Technologies

Richard L. Amoroso

ABSTRACT

A NEW 'AGE OF DISCOVERY' inclusive of the 'Physics of the Observer' (nature of awareness) – discovery of the mind or consciousness is pragmatically imminent. A comprehensive noetic theory only awaits completion of empirical tests that like the discovery of electricity leads to a plethora of technologies – 'Conscious' technologies. A 'noetic effect' transducing physically real Cartesian mind-body interaction principles inherent in Einstein's Unified Field (U_F) is required. Quantum computing (QC) designs modeled to include U_F mind-body interaction parameters, operate the systems. Discovery of the mind is briefly reviewed, salient technology summarized; then favorite 'spirit-based' medical applications are focused on.

INTRODUCTORY PRÉCIS – TRANSITION FROM THE MODERN AGE TO AN AGE OF AWARENESS

Our main effort in AOC I [1] was to outline the imminent discovery of mind (awareness) and describe a few aspects of that discovery related to the mind-body problem. In AOC II [2] we further embellished our delineation of the Anthropic Multiverse [3-5] and then concentrated on defining the physical cosmology of love or the fundamental basis of spiritual union and how that relates to the structure and phenomenology of the soul. In this essay after reviewing new developments related to the cosmology of mind [6-8] and outlining pertinent basic assumptions of Unified Field Mechanics (UFM) [4,5] we concentrate on describing several of the more interesting millennial technologies and summarize the nascent experimental path for achieving them. We emphasize that if the new Unified Field (U_F) action principle considered synonymous with the 'spirit of god' is not physically real, none of the millennial technologies described would be possible to engineer. Therefore, before describing anticipated wonders that might be myopically considered improbable such as telecerebroscopes, telepathy machines or sensory by-pass prosthesis where all the blind may see; we feel mandated to describe important seminal experiments to discover, isolate and mediate the new U_F action principle to help ameliorate as much as possible the initial skepticism that always arises in historical paradigm shifts. Imagine for example if I wrote about 60 years ago, that I was going to align two mirrors inside a gas filled tube more perfectly than anybody ever aligned mirrors before, I was going to put a ruby crystal between the two mirrors and wrap an ordinary photographic flash lamp around the tube. When turned on the device would create a light explosion so powerful it could burn a hole through steel! Crazy? Of course; and this is exactly what prominent physicists at the time said as reported by laser Nobelist Charles Townes: "*Several eminent physicists, among them Niels Bohr, John von Neumann, Isidor Rabi, Polykarp Kusch, and Llewellyn Thomas argued the maser* [laser precursor] *violated Heisenberg's <u>uncertainty principle</u> and hence* *could not work"* [9]. Cognitive scientists adamantly insisting Mind = Brain, like Lot's wife will be turned into crumbling pillars of salt.

Whether the spirit of God is physically real or not has been a profound historical mystery, not only for theology but also for philosophy of mind; especially cognitive science wherein a life principle is allegedly archaic and unnecessary. With the advent of the UFM paradigm, experiments can finally be performed to test this conundrum [8,10]. The scientific basis for this problem began ~400 hundred years ago when Rene Descartes' proposed a distinction between mind-stuff (*res cogitans*) and body-stuff (*res extensa*) [6]. The problem for contemporary scientists is that Descartes termed mind-stuff *immaterial* which was eagerly interpreted as *nonphysical* and essentially used to destroy credibility of interactive dualism. If *res cogitans* were nonphysical it would violate laws of thermodynamics and conservation of energy. But we know from usage of the term immaterial in the 17th Century that Descartes meant *spiritual* not nonphysical; thus, the path to UFM and development of Millennial Science is clear [3-6,11].

The term 'Modern Era', derived from the Latin *modo* (now) was coined to distinguish the transition from the 'Middle Ages' to the 'Age of Discovery' beginning with the Renaissance from about the 16th century. The last paradigm shift during our so-called modern era's age of discovery occurred at the turn of the 20th century with the advent of Quantum Mechanics (QM). It is reasonable to note that an imminent next shift is indicative of Millennial Science, an era deserving a new name related to the discovery and consequences of implementing aspects of consciousness. The term 'Age of Enlightenment' was already used for the 18th century. We would like to propose a purer nomenclature - 'Era' or 'Age of Awareness' in deference to the term *sapere verdere* used by Leonardo da Vinci meaning *"knowing how to see"* [12,13]. Isn't it time humanity evolved from *homo sapiens* (Latin: "wise man") to a new species *lucis sapiens* (wisdom of light) [14].

The term 'Millennial Science' is most probably acceptable by the theologically minded. The majority of contemporary Natural

Scientists (physicists especially) still struggling with the rift between science and theology created at the time of Galileo would typically prefer Unified Field Theory or Unified Field Mechanics (UFM); a penultimate step in the search for the TOE (Theory of Everything completing the Standard Model (SM) of particle physics and cosmology), albeit only the next convenient resting point in mankind's quest to understand the nature of reality or Holographic Anthropic Multiverse [3]. Curiously serendipitous, the anacronym TOE is reminiscent of toe-hold or foothold – *relatively insignificant inroad from which further evolution may occur.* UFM or millennial science takes us beyond the SMs virtual 3D reality we observe into higher dimensional (HD) space which is the limit of our local universe or 'creation' [15]. This TOE or toe-hold, because it reveals the spirit of God takes us to his 'foot stool' [16], but this is not the end – it is the door (Fig. 2) to a multiverse with '*room for an infinite number of nested Hubble spheres* (beyond our visible universe) each with their own fine-tuned laws of physics' [3] or theologically, room for other creations – "*like grains of sand at the seashore*" [3,17].

The current buzz in physics is that '*spacetime is not fundamental; it is emergent from unknown underlying new principles*'. To clarify from the history of human epistemology (nature, scope and theory of knowledge); our understanding of existence began in the dark ages of myth and superstition. Then the ancient Greeks birthed logic and reason. When logic failed in certain cases, (gravity for example) Galileo instituted experimental science which culminated in Classical or Newtonian Mechanics, called the 1st regime of natural science. The 2nd regime arose at the turn of the 20th century with the advent of QM. Now we imminently approach entry to a 3rd regime of UFM - The profound domain where physically real aspects of the *Spirit of God* or light of the mind resides! Entry to the 3rd regime also completes the tools of epistemology:

1) Myth and superstition
2) Logic and reason
3) Empiricism/experiment

4) Transcendence – Final tool of epistemology, all psi phenomena and revelation from God.

Transcendence breaks down the 1st person − 3rd person barrier allowing experience other minds firsthand [6-8,10], the principles of which, as shown below, are key criteria for the spirit mediated quantum computing (QC) parameters essential for conscious allopathic medical technologies.

Einstein said his relativities touched all of physics; but UFM will cause a tsunami touching everything to the extent of once again closing the rift between science and theology, such that they can no longer be considered mutually exclusive, but rather opposite ends of the long continuum of tools of human understanding, shaking, but not crumbling the lofty edifice of academia. This essay, briefly reviews basic tenets and applications of UFM, then focuses on the cornucopia of marvelous new 'millennial medical technologies' just around the corner.

EXTENDING REALITY - BEYOND THE VEIL OF SPACETIME

The 3rd regime of reality can be considered to a gated community having an impermeable gatekeeper, the quantum uncertainty principle. In fact, the uncertainty principle, one of the most fundamental premises of QM, experimentally 'proves' it is impossible to simultaneously measure both aspects of complementary quantum observables like position and momentum (or energy and time) with precision; the more accurately the 1st is measured the less accurate the measure of the 2nd. This wall of uncertainty has been inviolate for a century blocking access to the physical U_F domain revealing how the 'spirit of God' acts as a de Broglie–Bohm super quantum potential guiding the evolution of the 'contents' of 3-space. But before the discovery of QM, we had no idea there was an actual (scientific/theological) gate, or how it operated in the scheme of things. However, 'evolved' visionary people 'knew' a gating mechanism existed that allowed spiritual insight to enter the mind. Just as the tools of QM were invisible to classical mechanics, until now the tools of UFM have been invisible to contemporary methods of experimental science.

The author, a noeticist (physicist/priest) has learned to use 'transcendence' as a tool in theory formation [6,18,19]. Such 'gifts' of Millennial Science are expected to become more and more common among researchers [4,6]; not however meant to replace empiricism, rather to facilitate it. After hearing a voice of the spirit (1995 revelation): *"quantum mechanical uncertainty is the mystery, even the mystery of God"*; it took ten years of pondering to sufficiently develop an anthropic cosmology [3] that provided a proper fundamental basis for understanding how to surmount the uncertainty principle [4,5]; then another 4 to 7 years to design experiments [8,10].

Fig. 1 Amoroso

a) The veil of uncertainty provides an impenetrable gate demonstrated experimentally and locked tight for the last 100 years. b) UFM finds what's hidden behind the gate utilizing new tools.

Everyday flat Euclidean space our eyes observe has three infinite sized dimensions. The spatial dimensions of the semi-quantum manifold are limited in radius. A Euclidean observer peeking through the gate (Fig. 1) would see patterns of additional dimensionality in the 3rd regime; but an observer inside the 3rd regime could partially view the complete cyclical 12D structure of UFM reality because his vision would not be as limited by subtractive interferometry to the 3D limit of living systems. In noctic U_F theory, the 6D manifold of uncertainty (MOU) has an asymptotically finite radius that cyclically compactifies like an accordion (radius) that is also passed like a baton (unit brane segment) in a relativistic leap-frog (wave–particle duality) relay race. This complexity is not a violation of Occam's razor (simplest premises true) but represents the beauty of the gating mechanism separating the temporal manifold from eternity. If the gate was full open one would 'behold the face of God' and not be able to abide a temporal reality, so the complex structure rotates like a lighthouse beacon cyclically flashing into our biochemistry the life principle, our mind, the stream of qualia, and if into our 'eyes', psi or revelation from God – with good karma and meditation, the Lotus opens, as in Buddhist philosophy for example, to spiritual transcendence [6].

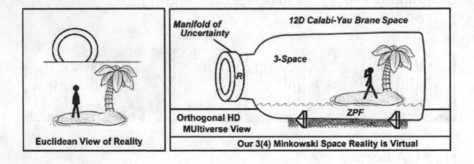

Fig. 2 Amoroso

a) The observed universe appears Euclidean and flat. b) But 3D reality is virtual ('created' that way just for the 'observer'). Beyond the finite radius Manifold of Uncertainty (MOU) UFM techniques discover large scale extra dimensions (LSXD) extended to infinity, nested in an infinite multiverse.

Let's take a moment to explain large-scale additional dimensionality (LSXD) more clearly; currently, a chaotic microscopic quantum foam of virtual particles winking in and out of existence is considered the 'basement of reality', a 4D manifold. Beyond the 'veil of uncertainty', matter as observed in 3-space (neutrons, protons, electrons - fermions) is comprised of unseen HD MOU, LSXD cyclic mirror symmetric brane components undergoing a duality of semi-quantum-ontological phase transitions revealing an extended UFM atomic structure beyond the current incomplete model of 3D Fermionic 0D singularities. Figure 3, attempts to illustrate the 1st rung only of this scenario. In Fig. 3a) the standard fermionic singularity is a knotted crossover or shadow of a richer HD brane topology hidden by the uncertainty principle by 4D experimental tools.

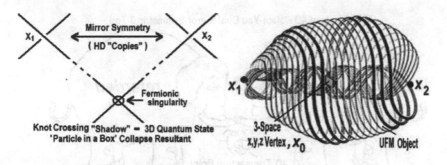

Fig. 3 Amoroso

a) Simplistic model representing new UFM view of matter. Only bottom X is observed in 3-space as a topological knotted shadow of HD standing-wave brane dynamics (notice the unknotted over - under crossings in top Xs). b) Leadbetter atom as a metaphor to illustrate the rich HD brane topology involved in the unseen structure of matter hidden in LSXD beyond the veil of uncertainty. The tiny central 3-space xyz vertex illustrates an expanded view of the crossing shadow or Fermionic vertex in a) plus second envelope.

The 3rd regime of UFM is highly symmetric, hinted at by Cramer's advanced-retarded, future-past standing wave transactional interpretation of QT [20] facilitating the continuous cycle of brane topology; the evolution of which is guided by a U_F 'force of coherence', a de Broglie-Bohm-like super-quantum potential synonymous with the spirit of God. Does the reader begins to appreciate the richness of this structure and why such complexity is required to mediate the temporal–eternal duality? With that said the HD mirror symmetry (Fig. 3a) is only the 1st duality; proper operation, requires two more involuted doublings (2nd in 3b). This hierarchical duality of mirrored crossings alone does not surmount uncertainty (open the gate). Only the 'mirror image of the mirror image' is causally free of the knotted shadow in 3–space. This key 3rd duality (not in Fig. 3) involutes like a trefoil, letting the crossing moves cyclically break closure like a 'butterfly valve' for noeon flux.

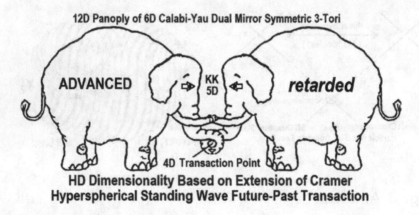

12D Panoply of 6D Calabi-Yau Dual Mirror Symmetric 3-Tori

ADVANCED — KK 5D — *retarded*

4D Transaction Point

HD Dimensionality Based on Extension of Cramer
Hyperspherical Standing Wave Future-Past Transaction

Fig. 4 Amoroso

Model of Cramer's transactional interpretation, showing symmetry and strength of the uncertainty principle held in place by the 'force of coherence' of hidden HD elephants. Figure adapted from [20].

NOETIC FIELD THEORY (NFT) – QUANTIFYING AWARENESS, GATEWAY TO MILLENNIAL SCIENCE

Millennial Science entails much more than the discovery of mind. Physicists have long balked at addressing the 'nature of the observer', academic psychologists daring to peer inside 'the black box' (consciousness) were not given tenure as recently as the 1990s;

discovery of the mind historically has been considered the oldest and most difficult problem facing human understanding.

The 3rd regime UFM, providing the tools/insights enabling the discovery of consciousness is also the arena where not only matter, but reality/existence itself can be manipulated as if it were 'Play-Doh' (clay-like modeling compound used in children's arts and crafts). All of my career I felt comprehending the mind-body problem would be the epitome of understanding; but now realize the 'fine structure constant' (FSC) pales even that, as it allows contemplation of the basis of existence within which 'awareness' is contained [21]. Needless to say, this makes access to the UFM arena potentially more dangerous than nuclear energy; but if we can behave ourselves, the wonders provided would bring life on Earth to the brink of utopia.

It is this author belief that 'all scientific discovery comes as revelation from God' [18,19,22], if true how does that affect the development timeline of new societal structure and technologies? I feel the noetic technologies I want to help introduce are inevitable; but my personal shortcomings and the inertial roadblocks so firmly placed by myopic nay-sayers play an indelible role...

> *Every discovery in science and art, that is really true and useful to mankind has been given by direct revelation from God, though but few acknowledge it. It has been given with a view to prepare the way for the ultimate triumph of truth, and the redemption of the earth from the power of sin and Satan* [18].

The noetic formalism, Noetic Field Theory (NFT): The Quantification of Mind, is a comprehensive experimentally testable solution to the mind–body problem [4,5,8,10]. It is a form of Cartesian interactive dualism, currently abandoned by cognitive theorists as archaic and untenable, erroneously deemed nonphysical, violating the laws of thermodynamics and conservation of energy. For the NFT formalism, we define 'spiritual matter' (*res cogitans*), as physically real and thus subject to engineering methods, albeit generally invisible to the untrained eye or current empirical tools of standard model 4D quantum field theory [6,7]. In the evolution of physics (fundament of all natural science) we imminently enter the 3rd regime of UFM within the realm of which access to 'spiritual matter' enables whole new classes of conscious and consciousness mediated technologies; all of which require various forms of QC to operate [23]. Physical realization of Millennial Science with its new U_F action principle breaks down the 1st person − 3rd person barrier, finally allowing understanding of all aspects of parapsychology: telepathy, clairvoyance and telekinesis. Psychology will no longer be an art, but become hard physical science and the design of devices like Telecerebroscopes and technologies like conscious quantum computer music will also be possible [24].

THE PHYSICAL BASIS OF QUALIA

Qualia, plural of quale, defined in philosophy of mind means the essence of awareness, a *qualitative feel* associated with subjective experience like redness for example. The problem of awareness is tantamount to the problem of Qualia. If experience has a specific subjective nature; what would exist if the viewpoint of the subjective observer were removed? The remaining properties are those detectable by other sentients, the actual physical processes themselves or states intrinsic to the experience of awareness. This changes the perspective of qualia to the form *"there is something it is like to undergo certain physical processes. If our idea of the physical ever expands to include mental phenomena, it will have to assign them an objective character"* [6,25]. These are questions the framework of an integrative Noetic Science answers theoretically and empirically, allowing the 1st person-3rd person barrier to be broken - A major step in implementing consciousness based technologies and extra-cellular sentience.

Standard definitions of qualia are obviously an inadequate construct describing the subjective character in only philosophical terms. In the physical sense of NFT: describing qualia from the objective sense distinguishes the phenomenology of qualia from the underlying ontological 'nonlocal noumenon' or physical existence of the fundamental *thing in itself*. NFT suggests that a comprehensive definition of qualia needs to be comprised of three component forms considered physically real because the UFM noetic unified field on which the NFT is based is physically real. The proposed triune basis of quale is as follows:

Type I. The Subjective - The *what it feels like* basis of awareness. Phenomenological mental states of the qualia of experience. (The current philosophical definition of qualia, Q-I)

Type II. The Objective - Physical basis of qualia independent of the *subjective feel* that could be stored or transferred to another entity breaking down the 1st person-3rd person barrier. Noumenal nonlocal U_F elements and related processes evanesce qualia by a form of superradiance, Q-II.

Type III. The Cosmological - SOLS just by being alive represent a Qualia substrate of the anthropic multiverse, acting as a 'blank slate' carrier (like a television set turned on but with no broadcast signal) from within which Q-II are modulated into a physically real Q-I of experience by a form of superradiance. Note: Q-III has sub-elements called *quanemes* addressed elsewhere [6].

SOUL – THE 'SPIRIT' AND THE BODY

It seems important to briefly discuss how the noetic model defines the soul. Firstly, *'the spirit and the body is the soul of man'* [26]. To expand on Descartes generalization, *res extensa* is composed of temporal biochemistry (classical/quantum) and *res cogitans* is comprised of two components – an elemental intelligence co-eternal with God and an atemporal *élan vital* synonymous with chi, ki, *prāna* or the spirit of God which is the dualistic noeon field interaction principle (quantum/unified field duality). This triune structure is called the psychosphere (3-space and HD-space) and comprises the total bound of individuality [6]. Within this psychosphere there is an active cognitive domain; whether or not the cognitive domain comprises the whole psychosphere waits for future experimental verification by noetic holography.

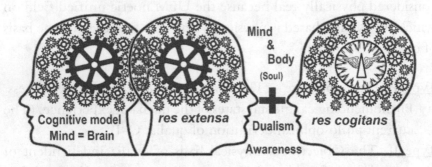

Fig. 5 Amoroso

a) Cognitive science claims mind = brain by biological mechanism (shown as gears). b) Dualism includes a soul and life principle. Mechanism is so efficient, and as an integral part of dualism; it is easy to see how cognitive science misses the need for the essential noetic element of awareness.

In addition, 3rd regime science magnificently completes the tools of human epistemology, which evolved from myth and superstition in the dark ages to logic and reason developed by the Greeks. When logic failed, Galileo introduced the current age of empiricism. The final tool epistemology is transcendence, once again reuniting science and theology, not as mutually exclusive, but is opposite ends of a wide continuum of human understanding.

THE PSYCHON – UNIT OF PHYSICAL MEASURE QUANTIFYING MENTATION

In order for Cartesian Interactive Dualism NFT, to finally take its place as the correct model of awareness, mental/life-principle energy parameters must be rigorously quantified in an experimentally accessible manner [8,10]. A new unit of physical measure, the 'Psychon', honoring Nobelist Sir John Eccles (discovery of synapse) is introduced to formally serve this purpose [7,27]. Eccles coined the term Psychon as a concept correlating awareness with a brain neural dendron (bundle of 100 apical dendrites with 100k synapses - 40 million dendrons in a human brain) [6].

In general, physics the energy of n-photons is $E = nh\nu = n\hbar(c / \lambda)$ with $\hbar$ Planck's constant and ν frequency. The second part of the equation is energy in terms of wavelength, l (in nanometers, nm) and speed of light, c. A unit of energy measure, honoring Einstein, quantifies one Mole (Avogadro's number, 6.02×10^{23}) of photons. Adaptation of the photon energy equation to measure Einsteins is similar, $E = N_0 h\nu = N_0 \hbar (c / \lambda)$ where energy of N_0 photons is instead in Einsteins, E. In photometry, one microeinstein, μE (millionth of an Einstein) per second per m^2 (m = meter) measures the power of photon em-radiation in photosynthesis imping leaf area. Because of conceptual similarity, we can use the Einstein as our starting point to create a unit of energy measure to quantify quale mental energy, called the Psychon and defined as one mole of 'noeons' - Putative U_F ontological exchange unit of mind-body interaction [6-8,27]. The term Noeon is derived from the Greek *nous*, mind and *noēsis / noētikos*, perception - what the nous does and the common "on" suffix in particle physics such as phot-on, exchange unit of the em-field.

Forces for three of the four known fields (electromagnetic, strong, weak) have experimentally verified phenomenological exchange quanta mediating their interactions by energy transfer. The graviton, and noeon, not yet discovered, according to NFT are not phenomenal field mediators as the regime of unification is not quantum but instead correlates with 3rd regime UFM topological parameters, with information exchange occurring ontologically (energyless) in HD brane dynamics, by a process called topological switching [3,6,23]. Sufficient delineation of ontological phase transitions is beyond the scope of this essay, but may be found in [4,5,23,27-31].

Mind-body Psychon coupling is not suggested to apply only to the noeon field impinging the synaptic processes in dendrons as Eccles proposed, but all arenas of mind-body interaction such as microtubules, neural nets, DNA or generally any biochemical spacetime interaction node within an entities Psychosphere [6]. To leave the reader of this volume with something, albeit somewhat simplistic, energyless topological switching occurs when staring at the central vertices of the ambiguous figure called the Necker Cube [23,32]. To actually utilize Psychon-noeon technology will require noetic holography briefly summarized below.

EXTENDING REALITY - BASIC EXPERIMENTAL PROTOCOL - ACCESS TO THE 3RD REGIME

The fermionic singularity or 'point particle' (0D object of no extension) used by physicists to explain most phenomenon mathematically for the last ~90 years; unfortunately, doesn't represent physical reality, only a way to model the 'location of a center of mass' for a hypothetical particle to introduce the magnitude as a variable in an equation. This is the state of the Standard Model (SM) of particle physics. String theory posits a resolution that elementary particles are 1D vibrating strings with a minimum size of the Planck length (10^{-33} cm); again, primarily a mathematical construct, with little physical significance other than black hole matter compression. M-Theory begins to do better with N-dimensional brane topology; but again, if these XD (called Kaluza-Klein) are SM, they must be compactified to the Planck length because there are not seen. UFM incorporates LSXD where the XD provide entry to the multiverse. The interpretation for why XD are unseen from 4D is called subtractive interferometry [3-8]. This is why a U_F solution to the point particle problem represents a paradigm shift; and what we are up against in designing an experimental protocol to enter the UFM 3rd regime [3,4,10,23].

As the first step, QM can no longer be accepted as an impenetrable 'basement of reality'. The existence of a LSXD 3rd regime is hidden by the uncertainty principle. We introduced various mind body-theological reasons above explaining why the new UFM parameters must appear complicated; essentially all of which address a new basis for a point as the entry vehicle [3-8]. Observed reality is defined from inside an infinite virtual 3-sphere (Fig. 2a); now we must add dynamical insight to Figs. 3 & 4 revealing the continuous-state cyclical brane topology separating time from eternity and 3-space from 12-space. The first addition is a mirror symmetric antispace from which real space emerges from the quantum vacuum [3-5]. We hope Figs. 6 & 7 give insight into this scenario. In Fig. 6, assume the i,j,k propellers represent the chaotic antispace from which periodic coherent modes (cube faces) couple to produce the Bloch 2-spheres at the bottom.

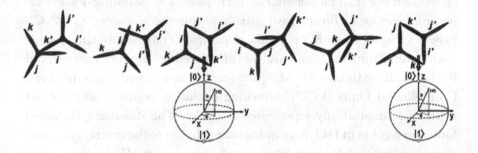

Fig. 6 Amoroso

Continuous-state mirror symmetric space-antispace vertex rotations, $i, j, k : i', j', k'$ (shown as propeller pairs) periodically couple (cube face) to form Euclidean 3-space resultants (shadow Fig. 3a), shown below as quantum Bloch 2-spheres. The little circles in Fig. 7b below also show the same vertices and cube faces. The 2-spheres form a line (locus of points) as the evolution of an x-axis for example.

Figure 7a illustrates more of the continuous-state process from the point of view of Cramer's standing-wave Transactional Interpretation of QM [20]. A present instant, moment when a Bloch 2-sphere arises (Fig. 6) occurs when an offer wave and confirmation wave entangle.

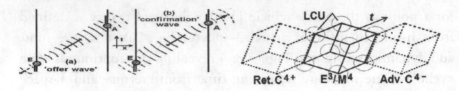

Fig. 7 Amoroso

a) A Cramer transaction realizes a present moment point, x,y,z,t as a standing-wave of future-past advanced-retarded offer-wave confirmation wave parameters. b) How observed (virtual) 3D reality (central cube) arises from the infinite potentia of 12D space. The 'standing-wave-like' (retarded-advanced future-past) mirror symmetric elements, C^{4+} / C^{4-} (where C^4 signifies an 8D potentia of complex continuous-state space is distinguished from the locally realized 3D spacetime) produce the observed Euclidian, E_3, Minkowski, M_4 space (center) as a closed resultant. Least Cosmological Units (LCU) governing evolution of the 'points' of 3D reality are simplistically represented as circles. The Advanced-Retarded future-past cubes in HD space guide the evolution of the central cube (our virtual reality) that emerges from potentia elements of HD U_F space. This is an extension of Cramer's model to HD space.

The mainstream physics community has marginalized the concept of a fundamental polarized Dirac vacuum because it is myopically believed to conflict with highly successful Gauge Theory. Nevertheless, our basic experiment in the Dirac context – interferes with the vacuum utilizing a laser oscillated rf-pulsed resonance hierarchy set up to interfere with the periodic (continuous-state) 'beat frequency' shown in Fig. 6 [4,8,10]. The experiment incursively 'pokes a hole in 4D Minkowski spacetime', revealing the hidden until now HD space between the Bloch 2-spheres, allowing U_F energy into a detector. Imagine (Fig. 6) that each of Bloch spheres at the bottom represent discrete frames of film (movie theater model); and the propellers a rotating relativistic field between them. We see how the chaotically spinning propellers (uncertainty) periodically form faces of a cube (observed reality). This occurs as a 'beat frequency', which because of the polarized nature of Dirac spacetime can be experimentally interfered with surmounting uncertainty.

For preliminarily predictions we could calculate hyperspherical volume of 4D and 5D space [10]. The general n-volume equation is $V(n,r) = \pi \frac{n}{2} r^n / \Gamma\left(\frac{n}{2}+1\right)$ where $V_{n,r}$ is volume per number of dimensions, n of radius r and Γ a factorial constant. These n-volume equations relate to predicted volumetric properties of the semi-quantum limit of a Manifold of Uncertainty (MOU) of finite radius, which is the intermediary between 2nd regime QM and 3rd regime UFM.

n		$V(n,1)$
0	0	0
1	2	2
2	π	≈ 3.1416
3	$4/3\pi$	4.1888
4	$1/2\pi^2$	4.9348
5	$8/15\pi^2$	5.2638
6	-	∞

Fig. 7 b. Table 1. Amoroso

TABLE 1. Increase in standard Hypervolume values as dimensionality, n increases for radius, r of a unit sphere or n-ball equal to 1. For the last 100+ years spectral lines have been measured in 3-space. The small added volumes for 4D and 5D can be used to calculate additional hidden HD spectral lines.

The purpose of the initial experiment is to discover XD beyond SM 4D, by predicting new spectral lines in hydrogen below the lowest Bohr orbit [10], making no sense in 3-space. Done by sending signals into the periodic 4D and 5D hypervolume cavities (TBL 1) by rf-pulsed resonance. The first two orbits in hydrogen are at .5 and 2Å; the experiment predicts hidden lines between .5 and 2Å, discovered if rf-pulses are reflected back from the HD cavity into a spectrometer. Once UFM XD are verified, with simple phase manipulations, energy of the U_F may instead be emitted from the LCU trans-dimensional gateway for use in Millennial Technologies.

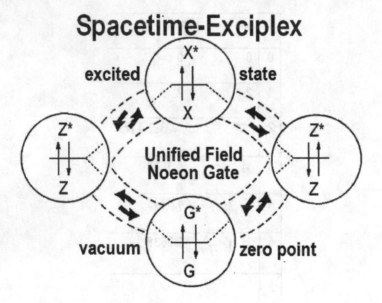

Spacetime-Exciplex

Fig. 8 Amoroso

a) Least cosmological unit (LCU) exciplex complex tiling the spacetime backcloth of multiverse cosmology. The array of LCUs acts as a gating mechanism for entry of the U_F into every point in Minkowski 4-space and all matter. Each of the four spheres represents one triune LCU complex (Fig. 9a). Imagine the two Z circles represent a virtual particle on an $x_1 \rightarrow x_2$ line element; it can only exist for the Planck time. An exciplex, short for excited complex, by circle X can remain in an excited state, providing a UFM mechanism for surmounting uncertainty for entry into the LSXD 3rd regime.

Quantum mechanical uncertainty has been empirically demonstrated as an inviolate law of nature; in fact, most quantum physicists believe its violation is impossible. This is absolutely true; but only as a limiting case in terms of the currently used Copenhagen Interpretation of quantum theory – a 4D model. A sparrow fits snuggly 'confined' inside the cardboard tube the size common paper towels come on; the birds' wings are pressed against its body (domain wall of the tube) and cannot be opened to fly. The sparrow needs additional space to open its wings; the principle is similar for opening the gate of uncertainty. In 4D quantum field theory a zero-point field (ZPF) virtual particle winks in and out of existence from the stochastic quantum 'foam for a brief period limited to the Planck time, $t_P \equiv \sqrt{\hbar G/c^5} \approx 5.39 \times 10^{-44}$ seconds (G gravity, c speed of light) as dictated by the uncertainty principle. The additional LSXD degrees of freedom supplied by UFM dynamics allow semi-quantum limit spacetime to be modeled as an exciplex LCU complex [23].

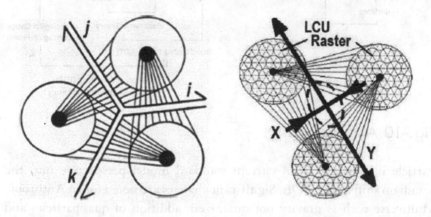

Fig. 9 Amoroso

Conceptual views of the structure of close-packed Least Cosmological Units (LCU) tessellating spacetime in the semi-quantum limit. a) An LCU complex structure comprised of a duality between the semi-quantum

MOU and LSXD brane transformations, thus the three i,j,k parallels represent a Witten 1D string vertex [3,23] a step above the 3-space knotted shadow (Fig. 3a). The underlying circles with field lines are a magnification of b) which shows the 3-space knotted shadow.

The Least Cosmological Unit (LCU) paralleling the 'unit cell' that builds up crystal structure is literally the 'key to everything', here a play on words for the TOE or Theory of Everything, a paradigm taking natural science into the next regime. Like quark confinement, a solitary LCU does not exist in nature but is a close-pack conglomerate of 12 LCU [23]. The close-packed exciplex model LCUs tessellate the MOU which can be rf-pulsed into various vacuum programming configurations.

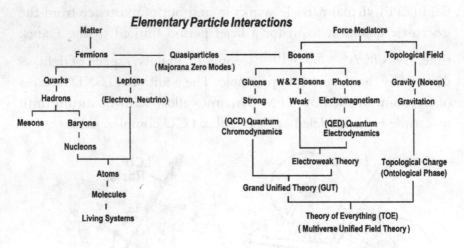

Fig. 10 Amoroso

Particle interactions from current standard model perspective into the paradigm shift to the TOE. Significant changes are noted for an Anthropic Multiverse such as gravity not quantized, addition of quasiparticles and Ontological-Phase Topological Field Theory (OPTFT) required for QC and awareness. Figure included in this essay because the UFM distinctions on far and bottom right are essential theoretical elements producing the paradigm shift to Millennial Science. The physics of reality is about the dynamics of matter through space, time and now eternity, and our existence and manipulation of it.

UNIVERSAL QUANTUM COMPUTING (UQC)

Since virtually all the wondrous applications of Millennial Science, mind-body, city infrastructure or otherwise, require some form of Universal Quantum Computing (UQC) [23], it behooves us to give a brief overview. The concept of Quantum Computing (QC) began in 1981 when Nobelist Richard Feynman queried if *'quantum physics could be simulated on a computer'*, quickly seen to require a new kind of computer called a *'quantum computer'*. Moore's law (Gordon Moore, Intel's founder), never wrong in over 50 years, states that since the invention of the integrated circuit in 1958, the number of transistors on a chip doubles about every two years (2,300 in 1971), as well as processing speed, accompanied by a corresponding reciprocal shrinkage in the size of a transistor. Right on schedule Intel ICs have 20 billion transistors in 2018, extrapolating to 100 billion (number of neurons in human brain) by 2020. Since lithography is at 7 nn using narrower uv (ultraviolet) techniques, commercial QCs will be available soon, because at that level electron transitions become discrete (quantum) rather than continuous (classical).

For standard computing a bit has two states, zero or one; but for quantum bits (qubit) the states entangle. Entanglement allows superposition of multiple quantum states between zero and one, such that only a few hundred qubits scale to quantum states equivalent to the number of atoms in the universe (10^{80}). Standard qubits are realized on the Bloch 2-sphere (bottom row Fig. 6). For UFM UQC, requisite additional 4D degrees of freedom are introduced by defining a relativistic basis for the qubit (r-qubit) in a higher dimensional (HD) context, providing the additional degrees of freedom for Relativistic Information Processing (RIP) [33-35].

In physics, theory bears little weight until supported by rigorous experimental confirmation, less if new, or proposes a radical paradigm shift. Our model, temporarily ahead of its time, claims UQC can only be achieved utilizing '3rd regime' Unified Field Mechanics (UFM). What distinguishes our work from myriad avenues to UQC? Virtually all R&D paths struggle with decoherence (quantum state destruction by interaction with environment). If the highly-favored room-sized cryogenically cooled (attempt to prevent/delay decoherence by ~absolute zero temperatures) QCs are ever implemented, they would be reminiscent of the 1946 city block-sized Eniac computer containing 17,468 vacuum tubes. The Uncertainty Principle and Decoherence do not apply to our room temperature, tabletop prototype [23], bypassing the intermediate 'Eniac step'. More dramatically UFM UQC is not limited to the confines of 'locality and unitarity' which are the fundamental bases of QM itself, as one might imagine if QM is left behind. Thus generally speaking, the UFM UQC model could putatively be implemented on any other viable quantum platform. Albeit, here's the rub: a conundrum still exists; new classes of QC algorithms designed for HD UFM brane topology remain to be developed.

The last age of discovery occurred about 100 years ago. Discovering the remaining requirements for UQC will usher in the next one. Scott Aaronson of MIT said, '*quantum computing requires a new discovery in physics, and if that discovery has been made is has not been revealed to the physics community.*' Gazing over the landscape of the QC research community, everyone seems to be about on the same page in battling decoherence as the final problem to implementation. With relative certainty, the discovery is revealed in our UFM model which naturally supervenes uncertainty and decoherence. The transition to UFM is not a single, but a 'bucket of discovery', perhaps similar to the ladling out of QM a hundred years ago, by the seventeen eventual Nobel laureates who attended the world changing *Conseil Solvay* conference series in Brussels.

Physicists still 'believing' in a quantum universe where the Planck scale is the 'basement of reality', adamantly proclaim the impossibility of violating the quantum uncertainty principle. Here is how renowned science fiction writer Isaac Asimov put it, "*You can't lick the uncertainty principle, man, any more than you can live on the sun, there are physical limits to what can be done."* [36]. Physical limits apply of course, but the new tools available to us by UFM enable 'sight' beyond the rigid barrier kept in place for the last 100 years by the uncertainty manifold.

To define UFM QC as simply as possible: In the 4-space of the SM, current QC's take the uncertainty principle as a given, I/O operations destroy the quantum state, and decoherence limits run time. This is the manifold we observe; but it is a virtual reality imbedded in a complex HD space. In an Anthropic Multiverse, the Euclidean 3-space we see is organized just for us as observers. The duality is required so our temporal existence can 'surf' as it were on the face of eternity or we could not exist in Earthly form. Since God cannot have his omnies (omniscience etc.) if limited by the uncertainty principle, in HD space matter has a different form.

Let's refer to Fig. 3, where a particle like a proton is a knotted shadow, the knot of being the uncertainty principle. Above we see the L & R mirror images in HD. Not shown in the figure is that to be causally free, there must be additional mirror images of that first set of mirror images. So, the matter we see is really comprised of a hierarchal system of interacting brane topology that reduces to 3-space. Since the Dirac vacuum (spacetime) is polarized it can be manipulated by rf-pulses as in Nuclear Magnetic Resonance (NMR). If the rf-pules are set up for destructive interference, a hole is punched in spacetime opening into the HD realm. A second pulse timed for when the gate is open is used for I/O operations without decoherence and without uncertainty.

There is another very important point: The HD topological brane hierarchy comprising the UFM view of matter is not static, but in continuous flux; this means, (in addition to causal separation) a new copy of the 3-space quantum state is continuously cycling into place at the 'top' of the hierarchy. A 1D standing-wave, R_1 & R_2 requires two points, like the standard metric line element, $dS^2 = x^2 + y^2 + z^2 - t^2$. There is just one set of 12D parameters. We have tried to make it clear that temporal SOLS cannot abide the eternal space; and that subtractive interferometry shields us from this confabulation. The manner in which the topology handles this conundrum is by dimensional transformation: In continuous-state cycling spatial dimensions are transformed to temporal and then to energy, $s \rightarrow t \rightarrow E$[3]. The beauty of God's creation: This continuous cycling is inherent in the HD structure of SOLS, and as such gives us half the QC for free!

Matter/fields are manipulated by these transformations: the Galilean (classical) and Lorentz-Poincaré (quantum-relativistic); as might be suspected, UFM, QC needs an additional set of transformations to operate. The new framework is described by what we call Ontological-Phase Topological Field Theory (OPTFT) [23,28-32], designed especially to handle mediation of the Unified Field, U_F. OPTFT, essential to UFM bulk UQC will take us far into the future; with it, one leaves polynomial and quadratic algorithmic speedup in the dust as it will allow the possibility developing 'instantaneous algorithms' (not yet imagined by the general QC research community considering only local parameters) by utilizing the full EPR aspects of nonlocal holography.

It will take time to understand how to develop the complex algorithms for manipulating the deep structure of reality for implementing 2nd generation QC utilizing Ballistic information processing of the cellular automata structure tessellating the deep reality backcloth; this will also lead to de Broglie matter-wave defense shields making nuclear ordnance obsolete. Whereas general QC utilizes various quantum states of matter and particles, vacuum programming is based on the LCU semi-quantum UFM topological structure of spacetime. It won't be called U_F computing instead, because the term QC will stick. Note: QC programming with matter/particles alone leaves out the anthropic parameters related to the technologies related to consciousness.

MILLENNIAL SCIENCE UFM AND CONSCIOUS TECHNOLOGIES

Since Millennial/Conscious Technologies all require access to 3rd regime reality, it appears they must also all rely on various forms of QC for accessibility. This does not apply to any new meditative applications inspired by the greater understanding of awareness; see for example the 'mind sharing technique' [2,37]. Dozens of Millennial technologies will arise evolving into multi-trillion dollar industries. Here is a partial list, some of which we elaborate on.

I. Millennial Technologies – Naked Unified Field

The initial first generation battery of Millennial technologies will rely only on superficial U_F parameters devoid of aspects of consciousness because a more complex form of QC modeled after the mind-body interface is required to utilize the noeon field – A conscious QC is not conscious (see sentient androids in part II), but modeled to include interactive principles of awareness. Here is a summary of some salient 1st – Gen Millennial technologies.

- **Quantum Cryptography – Military & International commerce:** QCs will easily crack Military codes and encryption in international commerce; suggesting initially QCs could be taken over by eminent domain until new QC encryption algorithms are developed. Breaking encryption cannot be done in a reasonable amount of time on a classical computer. In 2009, classical methods were able to discover the primes_within a 768-bit number, but it took almost two years and hundreds of computers. It is estimated that it would take 1,000 times longer to break a 1,024-bit encryption key, commonly used for online transactions. A large-scale QC could theoretically break 1,024-bit encryption very fast. Some leading internet companies are moving to 2,048-bit keys, but even those are thought to be vulnerable to rapid decryption with a bulk UQC. The US NSA fears the

99

implications for national security: *'The application of quantum technologies to encryption algorithms threatens to dramatically impact the US government's ability to both protect its communications and eavesdrop on the communications of foreign governments'.*

- **Pharmaceutical Development:** Billions of dollars and dozens of years will be saved in R&D by pharmaceutical companies utilizing enhanced biomolecular QC design technologies.

- **Facilitated Web/Database Search Routines:** Vast improvement in data amount/speed, superior to massive supercomputer 'thinking machines' like IBM's Watson beating Jeopardy star.

- **Dramatic Improvement in Software Development and Testing:** Microsoft will finally be able to release error-free versions from day one (Europeans especially won't hate them as much).

- **Traffic Control – Autonomous Vehicles and Tiered Flying Cars:** QCs will provide sufficient processing power for multitiered stacks of flying cars – Vertical Take Off and Landing (VTOLS) to permanently end city traffic congestion now only seen in sci-fi films like *"The 5th Element"*, *"The 6th Day"* or *"The Minority Report"*. QC power will also allow more than 20,000 vehicles per hour to land at city airports allowing commuting and practical living to currently 'valueless' land 500 miles away (see weather control).

Ultralight carbon–fiber composites that can absorb 6 to 12 times more crash energy per kilogram than steel and reduce weight of VTOLS 50% make them safe and inexpensive to operate. Toray, the world's largest maker of carbon fiber, announced a 30 billion factory in Nagoya to mass–produce carbon-fiber auto parts for Toyota, Nissan and others. Toyota's 1/X concept car is one third the weight of a Prius, uses half the fuel, yet has the same interior volume. Weight, as Henry Ford said, *"is advantageous only for steamrollers"*.

Prerequisite technology is already falling into place. Most modern cities currently have fully automated or semi-automated commuter trains, automated traffic signals with computerized synchrony to enhance traffic flow along main thoroughfares and arrays of traffic cameras. Onboard computers in motor vehicles are becoming equipped with collision avoidance sensors and ability to set off an alarm or automatically stay in the middle of a lane. Autonomous highway vehicle Tech will fully mature first (< 30 years), with VTOLs ready and waiting, (<50 years).

- **Image Manipulation, Databases and Invisibility:** Currently a Google image search works poorly for searches by image; added QC processing power also leads to several forms of invisibility Techs (already prototyped) by background image projection on a foreground surface.

- **Coherent Control – Transistor Lithography:** Because of the uncertainty principle, about a dozen lines are laid down chaotically insuring at least one good line; with coherently controlled lithography another order of magnitude improvement from multiple to single line etching occurs (2018 7 nm by ultraviolet lithography). Next, a phase shift from classical to quantum substrates for QC with atoms/molecules instead of ICs occurs, maybe finally ending Moore's Law.

- **Coherent Control – Fusion Reactors:** The US National Ignition Facility (NIF) builds miniature stars that will produce nearly unlimited energy as soon as the problem of field containment is solved. Nuclear Fusion, the opposite of current fission based power, where nuclei are torn apart, is a process compressing atoms until their nuclei fuse, releasing a huge amount of energy. For laser-based inertial confinement fusion (ICF) [38], a cylindrical *hohlraum* enclosing a spherical deuterium-tritium pellet is vaporized. Coherently controlled confinement utilizing a UFM based QC solves the confinement problem leading to nearly unlimited low-cost energy.

- **Perfect Weather Prediction and Control:** Slow steady improvement in weather forecasting over the past several decades occurred because of increasing improvements in computing power and data analysis by supercomputers. Forecasting is very complex requiring an understanding of clouds, aerosols, land surface processes, ocean dynamics and very likely needs to include high atmospheric 'sprites' [39]. With the advent of QCs 'perfect' weather forecasting becomes feasible because a QC can process however many data points are required for perfect prediction. This leads to the immediate possibility for implementing weather control technologies especially needed for dissipating hurricanes saving billions of dollars annually in damages. The 2005 US Atlantic hurricane season, the most active Atlantic season in recorded history, had widespread impact costing an estimated 3,913 deaths and record damage of $159.2 billion.

Five of the seven major hurricanes - Dennis, Emily, Katrina, Rita, and Wilma - caused most of the destruction. A record twenty-eight storms formed, with a record fifteen becoming hurricanes. Of these, seven became major hurricanes, a record-tying five became Category 4 and a record four reached Category 5 (sustained winds greater than 157 mph), the highest categorization for hurricanes on the Saffir-Simpson Hurricane Scale. The need for an addition of Category 6 for storms with winds greater than 174 or 180 mph was suggested.

For a sufficient number of satellites in geosynchronous orbit, combined with ground-based data points, a UQC has the ability to analyze the immense quantity of data to precisely predict the weather for where hurricane vortices will appear. A Cluster of small thunderstorms can merge and turn into a hurricane if water temperature rises to ~ 80° F. The Gulf Stream provides such warm water paths for hurricanes to follow (Hurricane Alley). For hurricane control, airborne infrared or a geosynchronous satellite array of em-lasers blast the developing vortex core into a 200-mile-long wave front turning the hurricane into a tropical storm doing no damage. Alternatively, the satellite array could change the direction of a hurricane by creating an alternate warm water path for the hurricane or lead it to a cold-water region where it would fizzle.

- **Programmable Matter:** SciFi classics like *The Day the Earth Stood Still* (1951) had a door/ramp to the saucer that sealed seamlessly as if it were a solid piece of metal. The US Department of Defense (DOD) advanced research division, DARPA already has a Programmable Matter division soliciting grant applications since 2009 to develop a 'programmable matter universal toolbox' with one bar of programmable metal that could shape-shift into different tool shapes on demand, instead of carrying a toolbox with specific items like screwdrivers and wrenches.

- **Antiballistic Defense Shields:** Practical de Broglie matter-wave defense shields causing obsolescence of nuclear weaponry is additionally able to protect soldiers with personal shields, and vehicles or buildings from standard ordnance and radiation. The existence of de Broglie matter-waves has been demonstrated in numerous ways; but with no applications. Kevlar bullet-proof vests cause severe pain when hit by a projectile. This problem disappears for matter-wave defense shields because UFM shielding also contains an insulating layer of quantum vacuum so impact forces never enter the space of the protected region. The initial challenge is understanding how to develop the more complex UFM QC algorithms to manipulate the deep structure of reality to ballistically program the cellular automata structure tessellating the spacetime backcloth.

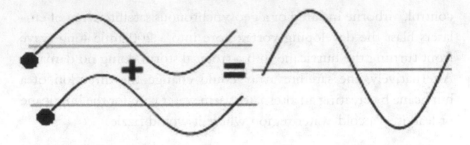

Fig. 11 Amoroso

Stones dropped in water create regions of constructive interference, or wave summation.

Utilizing a new concept of a static (albeit relativistic) de Broglie matter-waves, a resonance hierarchy can be used for constructive interference to summate matter-waves. According to Cramer's interpretation, matter is already a standing-wave (Fig. 3), thus with coherently controlled constructive interference of the inherent vacuum cellular automata, a prototype shield could be set up to strengthen aluminum to the density of depleted uranium (used in tanks). Operation of shield technologies relies on the central premise of extended de Broglie-Bohm-Cramer modeling of quantum theory wherein matter is continuously created, annihilated and recreated as physically real stationary standing-waves imbedded in the local fabric of spacetime. In HD space matter is comprised of mirror symmetric brane copies; summating these inherent matter-wave copies in conjunction with the arrow of time-eternity interface, and superimposing their topological phase gives the power factor for shield strength [23].

- **Energy Beam Weapons**: The US air force has developed reasonably mature em-beam (infrared laser) weapons; primitive in terms of 3rd regime UFM 'Star Trek' phasors, a 2nd generation of which could be configured as antimatter beams. As seen in the 'extending reality' sections near the beginning of this chapter, space and matter are much more detailed in HD than in the superficial view 3-space provides. As well-known, a laser beam is essentially a light explosion caused simplistically by 'perfect' mirror alignment. Likewise, superconducting occurs when what is called the 'mean free path' is unfettered by particle collision and 'ballistic transport' can occur in the superconducting medium. With UFM techniques cellular automata cavities tessellating spacetime can be programmed to 'borrow' the infinite energy of the polarized vacuum for many orders of magnitude more powerful beam weapons. Perhaps the reader now sees the overlap of UFM physics in some of the prior and ensuing millennial technologies.

- **SETI - Advent of the Q-Telescope:** If the Search for Extraterrestrial Intelligence (SETI) fails to access radio waves from technologically advanced civilizations for 50 years (~2030), likely to occur in a 'wrap around cosmology [40]; the search protocol will be reevaluated. A new type of SETI Q-Telescope is proposed to utilize properties of such an Anthropic Multiverse. This is required because our observed universe (3-space), closed and finite in time is causally separated from the rest of the infinite atemporal multiverse. The complex issues of creation/consciousness possibly thwarting SETI fulfillment is described in detail elsewhere [3-6].

- **Faster Than Light (FTL) Warp-Drive Interstellar Space Travel:** Superluminal space travel is no longer the stuff of Science Fiction; FTL development has begun! For the Alcubierre warp-drive metric (most advanced) derived from Einstein's General Relativity field equations by Miguel Alcubierre in 1994, space stretches in a wave contracting ahead of a ship and expanding behind it. Travelers inside the FTL warp-bubble move along what is called a 'freefall geodesic', with no local velocity so the infinite mass problem of special relativity is avoided. A major problem is that the model requires a Jupiter-size negative mass-energy to operate the warp-drive reactor, suggesting FTL space travel remains impractical for 1,000 years. The author's breakthrough Warp-Drive design, the 'Holographic Wormhole Drive' (HWD) [41,42] uses UFM physics to extend Alcubierre's seminal 'Warp-Drive Metric' [43-45]; bringing FTL drive theory to the brink of practicality by solving the serious energy problem plaguing other FTL designs.

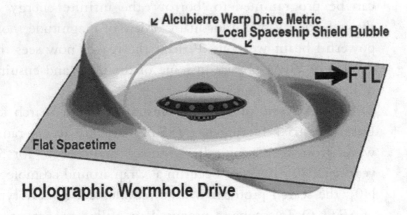

Holographic Wormhole Drive

Fig. 12 Amoroso

Holographic wormhole warp-drive metric using vacuum energy instead of a Jupiter-size negative-mass energy solves the FTL spaceship design problem.

There is a duality between Newton's (instantaneous, $v = 0$) and Einstein's (relativistic, $v = c$) theories of gravitation [46], hidden within the FTL concept itself that becomes more obvious when the HWD is understood in terms of the famous Einstein-Podolsky-Rosen (EPR) experiment that contrasts local finite velocities with nonlocal instantaneous entanglement. This duality is key to how the HWD operates and to understanding FTL 'Warp Factors'. It's a bit complicated of course, but this inherent duality: Newton-Einstein, wave-particle, local-nonlocal, mind-body, time-eternity, is a principle of cosmology arising within the nature of a point or singularity in spacetime. Physicists have ignored this very serious conundrum because it can be ignored in most calculations (or by 'tricks' like renormalization in quantum field theory).

The HWD uses what might simplistically be called a holographic figure-ground effect duality. The local Galilean velocity 'Warp-Bubble' is harmonically removed from the background of spacetime that embeds the usual 3D reality and drops it back in with a cyclic resonant beat frequency of LCU parameters. The greater the amplitude and duration of the wave, the larger the FTL background distance travelled between harmonic beats (wormhole jumps like 'skipping stones') - this is the warp factor. How to ballistically program Alcubierre's α and β vacuum functions to produce this effect is beyond the scope of this essay; but some technical details can be found the authors books and video presentations [40,41]. Essentially, unlike a photograph, a Hologram contains all information in any portion. Using the inherent dualities within a 'point' (LCU) by phase manipulated incursive oscillation of destructive (removal) and constructive (insertion) mirror symmetric interference, the Warp-bubble can be covered by an array of mini-wormholes that act in a leap frog process across the instantaneous nonlocal background. Thus, a Jupiter-size mass is unnecessary to bend space; instead optimized warping in conjunction with the oscillation of the phase geometry of the mini-wormhole oscillations is all that is required.

Actual implementation of FTL travel may be thwarted theologically in an Anthropic Multiverse where evolution is guided by the spirit of God; the reader will have to decide for himself. However, the author, Physicist/Mormon High Priest 'knows' the existence of the Judeo-Christian-Muslim deity. Although obscure, Mormon doctrine suggests a problem, not with developing FTL technology (same principles angels use to 'fly through the midst of heaven' without the passage of time, or Christ used to pass through walls when 'appearing in the midst' to the 12 apostles) but with actual use to visit other sapient civilizations: *"Only angels of the Earth are allowed to visit the Earth"* [47]. Sufficient hint that Nazi marauders like the Earth corporation causing havoc among the gentle Nauvi people on planet Pandora in the film Avatar, will not be allowed? Each world may have its own 'private; evolution; and when a technological society reaches the point of FTL travel their Millennium begins and they are taken off planet. After that spaceships are not required as transcendent 'heavenly beings' can see/experience the hologram ubiquitously. If this is Millennial Science in its purest form, it seems likely a 'do not touch' scenario will be in effect relative to interfering with other civilizations.

- **Manipulating Reality:** The initial realization that bulk UQC cannot occur without the utility of 3rd regime UFM came as a surprise; the scenario entails extending natural science beyond the 3-space of observation into the LSXD brane world. This is reminiscent of the 3-sphere visitor to Flatland [48] able to see the insides of the planar Flatlander's living embedded in a 2-plane. In that sense some aspects of access to the 3rd regime could become far more dangerous that nuclear energy because a UQC sophisticated enough to manipulate matter from within the nonlocal hologram of existence, potentially allows access and the ability manipulate any point in space with not only constructive, but also destructive interference even to the extent of the annihilation of some aspects of that arena!

II. Millennial Technologies – Mind-Body Noeon Interactions

2nd – Gen Millennial technologies go beyond 'naked' UQC in that the life principle inherent in the U_F is incorporated into the computing device. This represents a whole new class of QC modeled like a self-organized living system (SOLS) so that noeon flux may be utilized [6,23]. Utilizing UFM, new forms of noetic holographic techniques will see much deeper into the actual noetic structure of the 'soul' itself.

Noetic Holography

The MRI medical scanner utilizes quantum holography to glean information from biochemical atoms and complex molecules. The body is largely composed of water molecules, containing two hydrogen nuclei, or protons. Utilizing NMR techniques, hydrogen energy levels are excited (aligning with the direction of the applied field). Another magnetic field, the gradient field, is applied to kick the nuclei to higher magnetization; when the beam is turned off the protons return to the original alignment. Mapping the emission energy is used for diagnosis. Diseased tissue, such as tumors, can be detected because the protons in different tissues return to their equilibrium state at different rates. By applying additional variable fields specific slices can be selected, and images obtained by taking 2D Fourier transforms of the frequencies of the signal [49].

Functional magnetic resonance imaging (fMRI) is a neuroimaging procedure building on MRI scanning technology, contrasting the change in magnetization between oxygen-rich and oxygen-poor blood as its basic measure of brain and spinal activity by detecting associated changes in blood flow. This technique relies on the fact that cerebral blood flow and neuronal activation are coupled. When an area of the brain is in use, blood flow to that region also increases and oxygenated blood displaces deoxygenated blood rising to a peak before falling back to the original level. Oxygen is carried by hemoglobin. Deoxygenated hemoglobin (dHb) is more magnetic (paramagnetic) than oxygenated hemoglobin (Hb) which can be mapped by reconstructed cross sectional images of brain regions showing which neurons are active at a time [50].

Noetic holography is a dramatic step forward. While the two NMR techniques above only map atomic excitation in 3-space with no information about consciousness which is hidden behind the veil of uncertainty; noetic holography is able to take 12D cross sections of the structural-phenomenology of the soul. Firstly, SOLS have a triune structure [6]:

1) Eternal elemental intelligence, the fundamental bound of individuality,
2) The life principle, or spirit of God which is both the 'spark of life' and 'light of the mind, and
3) The physical body, which includes the brain which acts as a transducer of sensory data into the mind and processing unit for the metabolism and associated homeostasis.

We define the triune basis of SOLS as a hyperstructure domain called the psychosphere. The brain is not much more than a transducer of sensory and metabolic data and certainly consciousness (unified field) is not confined to that grapefruit size mass, but pervades the whole psychosphere. There is also a cognitive domain, but without experimental tests, we do not know if it encompasses all or part of the psychosphere. Like the spadix of the calla lily surrounded by the spathe, the Bible states: *"Even Solomon was not arrayed like one of these"*. We equate the body with the spadix and the spathe with the envelope of the noeon field.

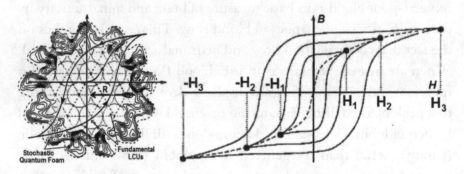

Fig. 13 Amoroso

Harmonic oscillation of noeon flux through the exciplex gate. a) Outer rung of stochastic uncertainty parted in center to reveal LCU tessellation. Noeon flux is cycled by a hysteresis loop. b) The gap size of exciplex gate hysteresis is a harmonic oscillation.

The process of noetic hyperspherical holography has similarity to fMRI quantum holography, but instead of creating cross sections by processing photon emission data from superficial valence electrons in neural tissue, harmonic oscillations of phase dynamics for topological charge are timed to collapse into cyclical nodes that emit soul noeons. There will be an 'alphabet' for qualia reminiscent of phonemes for sound, every disease will have a complex lattice-like topological structure; likewise, health and modes of intelligence will also have unique topological signatures.

- **Noetic Holographic Scanner Consciousness (Mind-Body) Research Platform:** Experimental access to the noetic life principle of the U_F as a holophote (lighthouse) 'beam' of noeons is like the discovery of electricity several hundred years ago. Initial research will experimentally settle the issue between Biological Mechanism – the premise of the Cognitive approach that mind = brain and Cartesian Dualism requiring an additional life principle [6-8]. Since the Cognitive approach is currently embraced by 95% of mind researchers; it has been a monkey wrench slowing the cogs of progress. 4D spacetime is sealed by the quantum uncertainty principle; HD brane space of UFM has sufficient parameters to allow manipulation of the noeon gating mechanism so that LOVERs [3,11] can be experimentally designed. This initial work will prepare the way for all the imminent mind-body conscious technologies.

- **Permanent End to World Hunger:** A permanent end to world hunger is now possible even for an unlikely inordinate world population of a trillion. For decades, vegetables have been grown indoors by a process called hydroponics, where a vat of nutrients bathes the roots and grow-lights produce photosynthesis. Recently, commercial protein culture such as fish cells on mesh screens has begun [51,52]. This technique however only grows food at standard growth rates. A UFM QC is not just fast but since it is able to manipulate reality itself, this includes focusing of the arrow of time for growing living tissue (artificial protein) at a vastly accelerated rate for food. The principles are similar to those outlined in the 'medical tricorder' bullet below.

The process is based on parameters inherent in an anthropic multiverse – a life principle that guides the evolution of SOLS. HD UFM reality reveals not only the nature of the arrow of time but also how to manipulate it by a process called constructive interference. Imagine throwing stones in a pool of water. Ripples flow outward in concentric circles. If one throws two stones near each other; in some areas, waves add together or summate called constructive interference. This occurs with a beat frequency like the revolving lens of a lighthouse beacon. This cycle is inherent in the backcloth of spacetime and may be manipulated for constructive interference of the arrow of time set up to focus the inherent 'eternity wave', $\aleph$ on the cell culture array consisting of a multilevel graphene layer which is both the QC substrate and protein cell mesh.

There is a lot of parameter overlap in these millennial technologies; here, summating the arrow of time has conceptual similarity to the warp-factor in FTL spaceships. For mankind to live in 'time', our temporal 4D universe must 'surf' as it were on the 12D 'face of eternity'. The additional 8D are invisible because of subtractive interferometry. The nonlocal hologram is instantaneously interconnected; summation of its Least Units (LCU) in various ways creates defence shields, FTL warp drives AND accelerated food growth!

- **Sentient Androids:** Enhanced Artificial Intelligence – Personal/Home Robotics. With a class of 'conscious' UQC modeled after the mind-body interface sentient or pseudo-sentient androids are possible. The mind-body interface is thought to be a naturally occurring form of 'conscious' quantum computer. By pseudo-sentience we mean the issue of conscious machines remains difficult compounded by the

'Chinese Room' experiment [53], suggesting it could also remain a challenge experimentally. The problem cannot be solved philosophically only laid bare to certain probabilities. Virtually infinite data bases and sophisticated AI algorithms should allow realization. What would we do with them is more challenging. Would humanity finally come to the point illustrated in a Star Trek episode, where being born entitled one to all basic necessities, housing, medical care, education, and resources. The challenge was self-improvement and work toward the betterment of mankind and environ.

What are the Criteria for Sentience? Sentience is suggested to be synonymous with an entity having subjective experiences also known in Philosophy of Mind as experiencing qualia. Sentience is often considered to be distinct from other aspects of mind like intelligence, self-awareness or free agency. It is possible to list salient components of consciousness. We suggest four: Sentience, Intelligence, Self-awareness and Free will. Must a conscious (self-aware?) system be considered alive? We have addressed this issue elsewhere in what we have termed System-Zero: The proteinaceous unit called the prion, (responsible for neurodegenerative encephalopathies) a particle 'below' the virus. System-Zero propagates from normal to infectious by a conformal change in the protein structure by action of the force of coherence of the U_F [4,6].

Following the assumptions:

1) A physically real noetic 'life principle' beyond 'biological mechanism' exists inherent in the action of the U_F,

2) The mind-body interface (Cartesian interactive dualism) is a class of naturally occurring 'conscious quantum computer' (not that the QC is conscious but modeled after such UFM principles) and

3) Combining the two concepts leads to truly sentient androids when utilizing the stated class of UQC systems modeled after the noetic or Cartesian mind-body interface.

The noetic QC Android model is empirically testable with experimental protocols summarized herein. Access to U_F action of the life principle requires surmounting the quantum uncertainty principle. Furthermore, the required bulk UQC cannot be achieved with 4-space parameters and requires M-Theoretic principles of UFM cast in LSXD. Because qualia are physically real our NFT allows breakdown of the 1st person barrier leading to the 'extracellular containment of natural intelligence. We believe implementing some form of sentient android devices is only this far away!

- **Conscious Quantum Computer Music (CQCM):** CQCM is music is any music genre integrated with qualia utilizing the breakdown of the 1st person-3rd person barrier such that 'digitized' qualia will provide a fantastic new form of entertainment and perhaps music therapy. Qualia will be embedded in the CQCM score; mood, or sensation will be physically transferred to the 'listeners' mind. CQCM is nothing at all like what is usually meant by computer music as electronic music. Notice how many UQC technologies overlap. Sensory bypass prosthesis, digital storage of qualia and telecerebroscope all intertwine.

Musicology would move beyond styles and instrumentation; A musician could compose a whole symphonic mentation of mood, sensation and feeling. This would require a new type of radio receiver if QC music were broadcast from CQC music stations. Just as radio receivers have volume, base, and other controls, a CQCM device would need a 'subtlety' dial for both individual tolerance and for setting how strong an emotion would be transferred to the subject's mind.

- **The Telecerebroscope:** A Telecerebroscope will allow thought to be recorded and shared with others as a subjective experience of visual qualia to the device user. This application can be considered a more advanced format of CQCM – mood, feelings to visual input. The Salvador Dali's of the world can make a fortune recording their dreams. Advanced models will allow one to peer out the eyes of a loved one in real time, or enjoy full blown telepathic communication. Telepathy/ Clairvoyant Machines or Telecerebroscopes breaking down the 1st person – 3rd person barrier by 'digitizing' the flow of consciousness (qualia) could allow mind-to-mind communication to become as common as cell phones. We have theorized that the quale of a 'pencil of light' is the pencil of light 6]; thus, the storage medium is an open question.

We have published elsewhere a shared breathing technique that allows one to share another's mind briefly, as if one were them [2]; so, from 1st hand experience we 'know' a mind-body modelled UQC has the potential of achieving such technology. For 4-space nonlocal EPR-like connectivity such a technology has been proven impossible by the no-cloning theorem [23]; but for the HD properties of UFM holographic space the situation changes. One sees the overlap in the techniques of these technologies.

III. Millennial Technologies – Mind-Body Interaction Medicine

These are the most advanced and therefore most complex Millennial UQC applications; built on scientific advances made by the prior two generations of UQC technologies. This group requires high level understanding of UFM spacetime programming. It includes noetic medicine and psychological or mind-body related applications. If the Cognitive approach (mind = brain) were valid, these 3rd Gen aps could have been relegated to prior generation UQC categories. They, to many, would be considered highly speculative applications; however, the main requirement, solving the mind-body problem, believe it or not, is already sufficiently complete, only awaiting experimental confirmation. Interestingly, the same experiments that will allow the implementation of bulk UQC, with minimal modification allow development of these phase three QC applications. Utility of UFM parameters is like the discovery of electricity; tens of thousands of phenomenal QC technologies await.

A new Noetic Effect is central to the mind-body interaction gating mechanism, balancing inertial entropy and syntropy for homeostasis (health or disease) [6,54]. An electromotive force (EMF) is induced in a conductor (copper for example) whenever it cuts through magnetic lines of force. In this regard, electrical generators have field windings surrounding a rotor or armature, also having windings. Rotation of the rotor within the field produces an electric current. In a simple case, a small hand cranked generator which uses U shaped horseshoe magnets to produce its field has its output connected across a large neon bulb. If the generator is cranked quickly, both halves of the bulb appear to be on. However, if the generator is run slowly, the two halves will alternately go on and off, showing that the output is AC. If there is no load (bulb) the crank turns with little friction, with a bulb connected the crank is hard to turn; this is the electromotive force. In terms of the duality inherent in the Noetic Effect, the action of which is centered on the geometric topology of the LCU exciplex gating mechanism; SOLS have complementary components similar to wave-particle duality, the temporal-life (particle) part is the stressor causing the electromotive force which if not constantly balanced causes any manner of disease conditions, and the eternal-spiritual (wave) part provides the instantaneous quenching or replenishing.

THE LOVER - LASER OSCILLATED VACUUM ENERGY RESONATOR

If the U_F is synonymous with an *élan vital*, life principle, ki, chi, *prāna* or 'spirit of God', allopathic medicine is about to receive incredible advances. Inroads have occurred with spirituality addressed in Holistic and so-called Conscious Medicine, but a real allopathic form of conscious medicine does not exist yet. Radical changes loom as bridges between science and spirituality improve. Proponents of spirituality in medicine explain that mind and body are inseparable and that emotions are crucial in health and disease [55]. Research on the transformative power of spiritual energy shows one can not only reverse disease processes and heal injuries but also attract more beneficial circumstances into one's life. Another author describes how new physics, with energy medicine and energy psychology, point the way towards a radical new approach to health and healing – one which is based on living in a conscious universe, rather than a material universe, tapping into the transformative power of spiritual energy [55,56].

Engineering access to the newly formalized action principle driving evolution of SOLS provides the means to address cause, prevention and cure of the approximately 400 known and heretofore incurable autoimmune disorders (as the spiritual root not discovered). For millennia, the three treasures of Traditional Chinese Medicine (TCM) Jing-Qi-Shen or essential energies taught in the Neijing, were the basis for Oriental health care; but with the advent of Allopathic or Scientific Medicine about a hundred years ago, TCM energy balancing techniques have been essentially marginalized and dozens of derogatory terms given for these 'so-called unscientific' treatment modalities. With the recent discovery of how to access and utilize the U_F of physics by a resonance hierarchy to coherently control (program) the exciplex gate in order to 'punch a hole' in what is called the Dirac covariant polarized vacuum, we have come full circle to provide the basis for a truly integrative allopathic medicine for the first time in history. TCM acupuncture is a primitive proof of Noetic Medicine. Although acupuncture meridians are in body tissue, tissue has a connection to nonlocal U_F energy (duality of noetic effect gating structure) as it is pervaded by noeon flux at the deep level.

Autoimmune Etiologies: The ~ 400 autoimmune conditions currently poorly understood and incurable are likely 'spiritual' conditions, whereby, psychosomatic or psychoneuroimmunological aspects contribute only superficially, which however, provide a framework for expanding the basis of autoimmune diseases and scientifically bringing the fundamental spiritual root into allopathic medicine [57]. The evolution of living systems as SOLS can be driven by external telergic factors regulating homeostasis. For our discussion, we use the autoimmune disorder, Alzheimer's Disease (AD) as a descriptive case. A significant advance, by Stefani was the realization that misfolds are a much more general property of proteins than previously suspected. Stefani's team also found evidence that 'the mortal blow' occurs just before the strand misfolds [58]

Recent discovery of ADDL protein (Amyloid ß-Derived Diffusible Ligand) allows the autoimmune trigger mechanism causing AD to be described by principles of complex systems theory. We have defined the new physical action principle as 'The Noetic Effect' and utilize the formalism of catastrophe theory to describe its phenomenology [59]. This Noetic Effect may drive the autopoietic system away from optimal equilibrium by catastrophes. This is the general noetic framework for utilizing the equilibrium surface of double-cusp catastrophe in terms of a hierarchical causation of potential jumps in state. In our example causing protein misfolds that trigger the onset of AD [59]. ADDL protein has been shown by several research teams to correlate with both Amyloid Plaques and Tau neurofibulary tangles [58]. Unfortunately, the 40 million spent on an ADDL vaccine was unsuccessful in preventing AD as people became ill in test trials.

The special class of UQC modeled after the UFM mind-body interface leads new classes of medical technologies able to ameliorate all autoimmune etiologies. Current treatment of autoimmune etiologies is typically by the utility of immunosuppressants which like most allopathic medical modalities does not address the root of the problem. The introduction of the UFM noetic field and its effects on protein conformation into an integrative medicine is vital to health [3-8,10,11,23,59]. This discovery entails a new action principle or force of coherence, inherent in the U_F driving the evolution of SOLS. UFM action additionally occurs on HD topological branes and is not limited to localized biophysical protein conformation; but the root of the force (noetic effect) causing negative conformation is in the topological charge of brane structure [1,41,60]. This is the panoply of the spiritual component for autoimmune conditions.

The author pondered long before creating an interesting term for the UFM 'vacuum transistor' switching the noetic U_F - Laser Oscillated Vacuum Energy Resonator (LOVER) [11], which instead of switching electron flow, mediates the noeon flux of the U_F. Won't it be a pertinent kick if the coming age of Millennial medical technologies are operated by a spacetime vacuum transistor called the LOVER! Twenty years ago, the author was 20 years ahead of the QC R&D community; at time of publication ~ 2018, as he continues to review the literature, they are about caught up, but only in terms of Topological QC with quasiparticle anyons in graphene [23]. My team could lose its lead at any time if this group sees how to access topological confinement utilizing phase doublings of zero mode Majorana fermions [28-31]. But my teams edge is not fully lost for two reasons: 1) The 2D anyon TQC will still require multimillion dollar cryogenics, 2) The actual discovery of a workable TQC will be somewhat unexpected in the manner it operates, and there will be virtually no UQC algorithms yet developed. Our model is tabletop and will cost pennies in comparison; but we don't have algorithms either which will have only partial overlap with TQC algorithms. We have been promised money; can we get at it in a timely manner?

Let's try to outline a predictive framework. As a physicist and an LDS High Priest, I utilize transcendence as a tool in theory development. In that respect, it is likely I will have a new set of key transformations summer 2017; but it is conceivable it could take 2 to 3 years to get experiment(s) done after which UQC can essentially be guaranteed. I humbly anticipate making final progress by revelation [6,18,19]. The rest of the QC R&D community could come to a reasonably similar place at any time by brute force of intellect and by trying numerous logical paths (also not devoid of inspiration). I mention this 'lunacy' because there is more going on here than an imminent immense paradigm shift to Millennial Science; Millennial Science not only completes the tools of epistemology (myth/superstition $\rightarrow$ logic/reason $\rightarrow$ empiricism) and finally to transcendence with a commonality tantamount to the other three tools; but also ushers in the actual Judeo-Christian-Muslim Millennium (if one accepts such a scenario). For all that has happened to me (and continues) I seem to be in the middle of it and expect 'God' has some statement to make by a priest rather than an atheistic physicist, or the *Zeitgeist* would not have the likes of me sitting in such a chair. This is an exciting time; one way or another UQC is inevitable.

God's hand is in all things. The doling out of scientific discovery (even though all by revelation – unconscious to most scientists) seems random from our perspective. There are a few times in history where God's hand is obvious; one is the 14-year-old *Jeanne D'Arc* (Joan of Arc) in the 15th century. France was about to be divided, ½ going to England. If so 400 years later instead of a transcontinental United States, there would have likely been (without taking the space to elaborate) 5 or 6 smaller nations formed in the region. The French were instrumental in the 'Colonies' winning independence from Britain. As it was the war against Germany was barely won; without a powerful America 'from sea to shining sea', it would not have happened…

So, let's make a guess based on best case scenario. I get a reasonable handle on the noetic transform by ~July 2017, sufficient to inspire my colleagues to prepare the experimental protocol(s). Experiment(s) performed by 2020; in interim algorithm development funded. Bulk UQCs by 2022 of some utility. Assuming QC is not taken over by eminent domain by governments because of cryptographic code breaking; 1st section application would appear in the latter 2020's. Please excuse if necessary my biopic commentary in these paragraphs. The primary reason is an 'empirical test' of prescience [6].

- **Sensory By-Pass Prosthesis (SBPP):** There are several SBPP applications; our favourite relates to vision. Currently retinal implant technology is improving (not to the extent of the mature artificial cochlear implant technology which operates better than natural hearing) to the degree that a blind test subject can find their way around recognizing doors and faces for example. Retinal implant technology will continue to evolve and improve with time. The major drawbacks are that retinal implants are only viable for the 14.5 % of patients with healthy optic nerves; and something about vision is more complicated retarding progress. A physical basis of qualia (awareness) is assumed. The brain is not the seat of awareness; but in the Cartesian interactive dualism model, a 'transducer of sensory data' from body (*res extensa*) into mind (*res cogitans*).

SBPP on the other hand, requires 'conscious quantum computing' [6,23]) taking advantage of the life-principle. One could say metaphorically that it operates resonantly like a tuning fork held over a partially filled glass of water vibrating the air. If the cavity provided by the empty portion of the glass has a volume complementary to the note "C" for example, a vibrating C tuning fork will make the water in the glass vibrate. This description is simplistic but the SBPP operates on the same principles. The string on a musical instrument is 1D, the skin on a drum 2D but vibrates into 3D space. Imagine a hyperdimensional 12D cavity and a carrier wave like those that induct 'sound' into AM and FM radio waves; but albeit in a much more complex manner. Common holograms create what we observe as a 3D image from a 2D acetate or glass plate (now also occurring electronically). Imagine instead a 12D object beyond the veil of spacetime (beyond our 3D observed reality) comprised of spacetime branes as boundary elements comprising a flowing series of perceptual (experienced) cavities in the mind; a mental hologram is created, not by an electromagnetic carrier wave as in radio but by a carrier-wave based on the life principle or spirit of God. The conscious QC acts as a transducer converting the external video image data and through a form of resonance induces the image train (qualia) directly into the mind.

- **Star Trek Medical 'Tricorder' – Diagnostic/Healing Device:** Does a Star Trek 'Medical Tricorder' able to heal a wound in a few moments that would typically take months under normal circumstances seem far-fetched; well believe-it-or-not, a true tricorder is only slightly more complex than growing protein at an accelerated rate by constructive interference of the arrow of time. While growing food by constructive interference of the arrow of time, is straight forward in comparison; the Tricorder is not much more complex. When used in conjunction with the noetic holographic scanner, with a database – genetic ailments, cancers etc. can also be healed. Once that level of technology is developed; some steps later would have the ability to 'beam' in (like the matter transporter) a piece of damaged aorta without invasive surgery.

- **Psychology Becomes a Hard Science:** Although psychiatry/psychology utilizes the scientific method in some aspects of testing and performance, psychology itself has remained an art. The severest conditions, personality disorders have remained incurable; at best patients are given an often-permanent battery of pharma and 'taught' with typically decades of therapy and life-long follow up counselling, some management skills to deal with the condition without needing hospitalization. Solving the mind-body problem in conjunction with UQC disease/person specific algorithms will finally provide remission with 'wearable' UQCs. Restructuring the soul must be individually tailored because no psychological/personality disorder is unitary, but comprised of not only numerous factors but often an overlap of multiple disorders.

Society generally allows citizens to walk around being 'floridly psychotic' as long as they are not a perceived danger to themselves or those around them. This of course includes being able to take care of one's self. This was clear in the extreme in the 2001 American biographical drama film 'A Beautiful Mind' based on the life of John Nash, Nobel Laureate in Economics. The film showed Nash as an extreme schizophrenic with hallucinations so real he could not distinguish reality from hallucination. Nash was a rare case in that with the help of a very patient and loving spouse he eventually learned to ignore his hallucinations and lead a semblance of a normal life.

With bulk UQC psychiatry and psychology will become a hard–physical science. The noetic holographic scanner will be able to fully diagnose the structure of the mind/soul in great detail and create multilevel cross section maps of any qualia and disease condition. This cannot be done with current fMRI techniques because MRI only accesses atomic energy levels and not 3rd regime brane dynamics mediated by the noeon field. The extreme complexity of the mind and overlap of multiple conditions requires significant noetic research. Once a database of mental disorders is created, through wearable electronics a specialized QC tailored to the individual will be worn. This is possible because the mind is a physical field like the electromagnetic field in many ways. Those who remember electron gun CRT televisions, may know that a fist size horseshoe magnet could bend and distort the image on the screen; and it didn't take much to permanently ruin the TV because the vanes in the electron gun were quite delicate and easily 'bent'. So simplistically put, a therapeutic QC would have the ability to reshape the mind to a more normal geometry.

REFERENCES

[1] Amoroso, R.L. (2012) Through the looking glass: Discovering the cosmology of mind with implications for medicine, psychology and spirituality, in I. Fredriksson (ed.) Aspects of Consciousness: Essays on Physics, Death, and the Mind, Jefferson: McFarland Publishers.

[2] Amoroso, R.L. (2014) What is love? The physical cosmology of spiritual union, in I. Fredriksson (ed.) The Mysteries of Consciousness: Essays on Spacetime, Evolution and Well–Being, Jefferson: McFarland.

[3] Amoroso, R.L. & Rauscher. E.A. (2009) The Holographic Anthropic Multiverse: Formalizing the Complex Geometry of Reality, Singapore: World Scientific.

[4] Amoroso, R.L., Kauffman, L.H. & Rowlands, P. (eds.) (2013) The Physics of Reality: Space, Time, Matter, Cosmos, Hackensack: World Scientific.

[5] Amoroso, R.L., Kauffman, L.H. & Rowlands, P. (eds.) (2015) Unified Field Mechanics: Natural Science Beyond the Veil of Spacetime, Hackensack: World Scientific.

[6] Amoroso, R.L. (2010) (ed.) Complementarity of Mind and Body: Realizing the Dream of Descartes, Einstein and Eccles, New York: Nova Science.

[7] Amoroso, R.L. & Di Biase F. (2013) Crossing the psycho-physical bridge: elucidating the objective character of experience, Journal of Consciousness Exploration & Research, 4:9; pp. 932-954.

[8] Amoroso, R.L. (2013) Empirical protocols for mediating long-range coherence in biological systems, Journal of Consciousness Exploration & Research, 4:9; pp. 955-976.

[9] Townes, Charles H. (1999) How the Laser Happened: Adventures of a Scientist, Oxford: Oxford University Press. pp. 69-70.

[10] Amoroso, R.L. & Vigier, J-P (2013) Evidencing 'tight bound states' in the hydrogen atom: Empirical manipulation of large-scale XD in violation of QED, pp. 254-272, in The Physics of Reality: Space, Time, Matter, Cosmos, Singapore, World Scientific; http://vixra.org/pdf/1305.0053v2.pdf.

[11] Amoroso, R.L. (2012) Spacetime energy resonator: a transistor of complex Dirac polarized vacuum topology, US Patent, http://www.google.com/patents/US20120075682.

[12] Heydenreich, L.W. (2016) Leonardo da Vinci, Italian artist, engineer, and scientist, Encyclopedia Britannica, https://global.britannica.com/biography/Leonardo-da-Vinci.

[13] Richter, J.P. (ed.) (1883) Leonardo da Vinci (Vol. 1) S. Low, Marston, Searle & Rivington.

[14] Amoroso, R.L. Lucis Sapiens - An Evolutionary Transmutation of Humanity: Treatise on the Nature of the Soul (in preparation).

[15] Smith, J. (1986) Book of Moses 1:33, *And worlds without number have I created,* Salt Lake City: Intellectual Reserve.

[16] Holy Bible(KJV), Acts 7:49, Isaiah 66:1, Matthew 5:35, *Heaven is my throne, and the earth my footstool.*

[17] Holy Bible (KJV) Genesis 22:17, *and as the sand which is upon the sea shore.*

[18] Young, B. (1946) Discourses of Brigham Young, J.A. Widtsoe (ed.) Salt Lake City: Deseret Book.

[19] Teachings of Presidents of The Church, Brigham Young (1997) Salt Lake City: Intellectual Reserve, The Church of Jesus Christ of Latter-day Saints; https://www.lds.org/bc/content/shared/content/ english/pdf/language-materials/35554_eng.pdf?lang=eng.

[20] Cramer, J.G. (1986) The transactional interpretation of quantum mechanics, Reviews of Modern Physics, 58(3), 647-63; (2006) https://www.slideshare.net/cliffordstone/cramer-john1.

[21] Amoroso, R.L. (2016) Gödelizng fine structure: gateway to comprehending the penultimate nature of reality, http://anpa.onl/pdf/S36/amoroso.pdf; http://vixra.org/pdf/1702.0176v1.pdf.

[22] Holy Bible (KJV) James 1:17, *Every good gift and every perfect gift is from above, and cometh down from the Father of lights, with whom is no variableness, neither shadow of turning.*

[23] Amoroso, R.L. (2017) Universal Quantum Computing: Supervening Decoherence, Surmounting Uncertainty, Hackensack: World Scientific.

[24] Amoroso, R.L. (2017) Intro to the genres and musicology of 'conscious' quantum computer music, NIME 2017, Copenhagen.

[25] Nagel, T. (1974) What's it like to be a bat?, Philosophical Rev., 83, pp. 435-450.

[26] Smith, J. (1893) D&C 88: 15, Salt Lake City: Intellectual Reserve, The Church of Jesus Christ of Latter-day Saints; https://www.lds.org/scriptures/dc-testament/dc/88.15.

[27] Amoroso, R.L. (2015) Toward a pragmatic science of mind, Quantum Biosystems, Vol 6, Issue 1, pp. 99-114, http://www.quantumbiosystems.org/admin/files/QBS%206%20(1)%2099-114.pdf.

[28] R.L. Amoroso, L.H. Kauffman & P. Rowlands (eds.) (2017) Unified Field Mechanics II, Singapore: World Scientific.

[29] Amoroso, R.L. (2017) Ontological-phase topological field theory: Einstein/Newton duality, Moscow, Proceedings PIRT 2017.

[30] Amoroso, R.L. (2017) The least cosmological unit: key to everything, ANPA 38, 2017, http://anpa.onl/.

[31] Amoroso, R.L. (2016) Ontological-Phase Topological Field Theory vixra.org/pdf/1608.0003v2.pdf; https://www.researchgate.net/publication/305769215.

[32] Ambiguous figure illusion (2017) https://www.google.com/search?site=&tbm=isch&source=hp&biw=1536&bih=735&q=necker+cube+illusion&oq=necker+cube&gs_l=img.1.1.017.2956.7361.0.9808.12.10.0.2.2.0.374.1115.6j2j0j1.9.0....0...1ac.1.64.img..1.11.1117.0..0i8i30k1.Hxw8OSM_Hzg.

[33] Vlasov, A.Y. (1999) Quantum theory of computation and relativistic physics, PhysComp96 Workshop, Boston MA, 22-24 Nov 1996, arXiv: quant-ph/ 9701027v4.

[34] Terno, D.R. (2006) Introduction to relativistic quantum information, arXiv:quant-ph/0508049v2.

[35] Peres, A. & Terno, D.R. (2003) Quantum information and relativity theory, arXiv:quant-ph/0212023v2.

[36] Asimov, I. (1989) The dead past, in Hartwell, D.G. (ed.) World Treas. Science Fiction, Little Brown.

[37] Amoroso, R.L. (2016) The physical basis of love or spiritual union, Vol. 7, No.7, Scientific God Journal; http://vixra.org/pdf/1511.0162v1.pdf.

[38] National Ignition Facility [ICF] Laser-based inertial confinement fusion, https://en.wikipedia.org/ wiki/ Inertial_confinement_fusion.

[39] High atmospheric sprites, https://www.youtube.com/ watch?v=0uo4nZtxMow; https://www.youtube.com/watch? v=brh—gYjZts.

[40] Luminet, J-P. (2008) The wraparound universe. Wellesley: AK Peters.

[41] Rauscher, E.A., & Amoroso, R.L. (2011) Orbiting the Moons of Pluto: Complex Solutions to the Einstein, Maxwell, Schrödinger, and Dirac Equations, Singapore: World Scientific.

[42] Amoroso, R.L. (2010) YouTube Links to Society for Scientific Exploration (SSE) 2010 keynote address on Warp Drive technology: The Holographic Wormhole Drive: http:// www.youtube.com/watch?v=ynxD-BaDxRk; http://www. scientificexploration.org/talks/ 29th_annual/29th_annual _amoroso _faster_than_light_warp-drive.html.

[43] Alcubierre, M. (1994) The warp drive: hyper-fast travel within general relativity, Classical and Quantum Gravity, 11(5), L73.

[44] Lobo, F.S., & Visser, M. (2004) Fundamental limitations on 'warp drive' spacetimes, Classical and Quantum Gravity, 21(24), 5871.

[45] Lobo, F.S. (2007) Exotic solutions in General Relativity: Traversable wormholes and 'warp drive' spacetimes, arXiv preprint arXiv:0710.4474.

[46] Amoroso, R.L. (2013) Unified geometrodynamics: A complementarity of Newton's and Einstein's gravity, in The Physics of Reality: Space, Time, Matter, Cosmos, pp. 152-163): London: World Sci.

[47] Smith, J (1974) The Doctrine and Covenants, D&C 130:18-19, Salt Lake City: The Church of Jesus Christ of Latter-Day Saints.

[48] Abbott, E. A. (2006). Flatland: A Romance of Many Dimensions. Oxford: OUP.

[49] Schempp, W. (1998) Quantum holography and magnetic resonance tomography: an ensemble quantum computing approach, Taiwanese Journal of Mathematics Vol. 2, No. 3, pp. 257-285.

[50] Pribram, K.H. (1999) Quantum holography: Is it relevant to brain function?, Information Sciences 115, 97-102.

[51] Stephens, N. (2013) Growing meat in laboratories: The promise, ontology, and ethical boundary-work of using muscle cells to make food, Configurations, 21(2), 159-181.

[52] Chiles, R.M. (2013) If they come, we will build it: in vitro meat and the discursive struggle over future agrofood expectations, Agriculture and Human Values, 30(4), 511-523.

[53] Searle, J.R. (1982) The Chinese room revisited, Behavioral and brain sciences, 5(02), 345-348.

[54] Amoroso, R.L. (2015) The physical teleology of syntropic interaction: an experimentally testable anthropic cosmology, pp. 1-17, http://vixra.org/pdf/1309.0039v1.pdf.

[55] Byrd, R.C. (1988) Positive therapeutic effect of intercessory prayer in a coronary care population, Southern Medical Journal 81 (7):826-829.

[56] Edwards, G. (2010) Conscious Medicine: Creating Health and Well-being in a Conscious Universe. London: Little Brown.

[57] Amoroso, R.L. & Amoroso, P.J. (2006) Elucidating the trigger of Alzheimer's disease: a complex anticipatory systems approach, International J of Computing Anticipatory Systems, D.M. Dubois (ed.) CHAOS, Liege, Belgium.

[58] Chiti, F., Taddei, N., Stefani, M., Dobson, C.M., & Ramponi, G. (2001) Reduction of the amyloidogenicity of a protein by specific binding of ligands to the native conformation, Protein Science, Vol. 10, Iss. 4, pp 879–886.

[59] Amoroso, R.L. (2004) Application of double-cusp catastrophe theory to the physical evolution of qualia: Implications for paradigm shift in medicine and psychology, in G.E. Lasker & D.M. Dubois (eds.) Anticipative & Predictive Models in Systems Science, Liege: CHAOS.

[60] Amoroso, R.L. & Chu, M-Y J. (2013) Empirical mediation of the primary mechanism initiating protein *conformation* in *prion* propagation, http://vixra.org/pdf/1305.0090v1.pdf.

THE ORIGIN AND HISTORY OF THE HUMAN MIND

by
Carl Johan Calleman

INTRODUCTION

THE NATURE OF CONSCIOUSNESS and the relationship between consciousness and matter have been discussed for a very long time. In this debate, there has been two conflicting schools of thought; the materialist and the idealist. The materialists have held the viewpoint that everything that exists is fundamentally matter and that the workings of the human mind can be understood from chemico-biological processes in the brain. On the other side of the fence have been the idealists who at its extreme end have held the viewpoint that physical reality is merely an illusion and that all matter is an expression of consciousness. In addition, there is the so-called dualist viewpoint, which sees matter and consciousness as two separate and yet interacting aspects of the universe. Dualism was notably elaborated by Descartes at the beginning of the scientific revolution in the early 1600's and is most likely the predominant viewpoint today.

It seems however that the discussion regarding the nature of consciousness is one that is intensifying at the current time, which indicates that we may again be in the midst of a paradigm shift that makes us look at reality in a new and different way. But why would this be happening now? And why was the dualist philosophy originally underlying the scientific revolution emerging in the early 1600's? Are these paradigm shifts something that "just happen" or is there some kind of underlying time line that determines when they take place? Modern historical research looks upon historical events as if they have no ultimate cause, while the ancient peoples of our planet, often within a "religious" context, looked upon human life as subjected to some kind of time plan, where different ages, eras or "worlds" replace each other in a predetermined way. Which viewpoint is right? Is history just a series of accidents or is there a pre-set time-plan for the evolution of consciousness.

In this article I will approach the nature of consciousness from such a historical perspective in accordance with the idea of a pre-existing time-line. Compared to abstract philosophical arguments such an approach has the advantage that it is based on empirical evidence, and so in a scientific way allows for a resolution to the question of the origin of consciousness. Empirical evidence correlating the various biological and mental eras in the history of our planet with the ancient Mayan calendar system has been presented in my books: *Solving the Greatest Mystery of Our Time: The Mayan Calendar* (2001), *The Mayan Calendar and the Transformation of Consciousness* (2004), *The Purposeful Universe* (2009) and most recently *The Global Mind and the Rise of Civilization* (2014). Such a historical approach places the evolution of consciousness and of human life in a context that is highly meaningful when it comes to understanding the purpose of life. The evolution of consciousness is however essentially a direct function of the evolution of the human mind. For this reason I will here focus on the mind, which is a more tangible and verifiable concept although also this ultimately is metaphysical in nature. The mind is something we cannot experience with our senses– we cannot taste, smell, hear or see it. The mind is here defined as "that which compartmentalizes the human perception and recreation of reality." We may learn about this compartmentalization and how it has created human intelligence from a historic analysis of events taking place when the mind has shifted. To be meaningful such an analysis will then need to be based on the Mayan calendar, which is a description of the timeline for the evolution of the global mind.

Overall it can however probably be said that researchers active in consciousness studies often have assumed that the relationship between consciousness and matter, or mind and the brain, is something that does not have a history and so has remained constant over time. This assumption, I assert, is not valid and once the precision of the time-plan for the evolution of the global mind is realized a new avenue for understanding the nature and history of the human mind opens up. The global mind drives human thinking, but would in fact evolve even if there were no humans on our planet.

THE SCIENTIFIC REVOLUTION

To exemplify this reasoning we may study a known time of activation of the global mind. We may then notice that the modern scientific world view and our present educational system does not recognize any theory (outside of the field of mathematics) that was formulated before 1617. No current-day school teacher would present any aspect of the medieval world view as true science. The oldest knowledge, which would be considered as accurate and useful science are Johannes Kepler's Three Laws of planetary motion published in his work *de Harmonice Mundi* from 1619. Kepler's Third Law may be considered as the first mathematical law of nature and is still used by astronomers to calculate the orbits of celestial bodies in our solar system. At around the same time, Galileo built his telescope and from his observations of the moons of Jupiter he argued that Copernicus' heliocentric worldview was correct. Yet, as late as 1616 the Inquisition had assembled a committee of theologians, who delivered their unanimous report condemning heliocentrism as "foolish and absurd". Nonetheless, over time the evidence from actual observations of the planets would gradually come to predominate and the geocentric view had to yield.

René Descartes, who was the first person to propose that our solar system was a product of a cosmic evolution process entered the scene at about the same time with his *Méditations* and *Sur la Méthode* (1648) that formulated the philosophical ideal of modern science that has come to go by the name of rational empiricism. Around the same time there was also an explosion of new mathematical thinking and development of more advanced computation techniques such as logarithm tables (Briggs 1617), which were instrumental for Kepler's calculations to support his Third Law. In the following years Descartes introduced Analytical Geometry, Fermat Infinitesimal calculus, and Pascal probability theory as well as calculation machines. The real novelty of the era was however not mathematics per se, but that it was applied to the natural world. This new paradigm led up to Newton's great synthesis of mechanics was published in *Principia Mathematica* in 1687.

The way historians so far have looked upon the scientific revolution and the Kepler–Galileo of mathematics to physics after 1617 has been to say that "it just happened", "people were freeing themselves from the constraints of the church and started to think in new ways." These are of course correct descriptions of what happened, but they are superficial and do not explain what made people think in new ways to begin with. After all, this explosion of novel mathematical thinking, and the even more radical notion that mathematics could be used to study the physical reality, must somehow have been related to a shift in how the human mind was used on a collective scale. Why else would this new paradigm emerge simultaneously in many different countries with researchers, who at the time had very little contact amongst themselves? Could this paradigm shift in effect have been the beginning of an activation of a global mind affecting the ways of thinking all over the planet?

Also we have reasons to ask why there would such a sharp shift in human thinking around 1617 so that modern science essentially accepts none of the physics before this year, but does incorporate Kepler's three laws published in 1619 as part of its accepted body of knowledge. Leaving aside the huge mentality shift that resulted from the contemporary Thirty Years' War and its conclusion at the Westphalian Peace Treaty and many other things that went on at the same time, it thus seems that there then was a very dramatic change taking place in human thinking. This led to the emergence of what we call science in the modern sense.

For the most part historians have not addressed what might have caused this significant shift in mentality around 1617. Historians of science have been content with noticing that "it just happened" and have not even tried to go into the issue of why it happened and why it happened at this particular time. And yet, it seems that at least Europe at the time was beginning the shift into a rational mind of a kind that had never been present before. Over the following centuries this way of approaching physical reality would spread and later become totally dominating – at least until the end of the 20th century. Naturally, if people were independently starting to adopt a rational mind at about the same time in the early 17th century the question deserves to be asked if this rational mind was truly a product of many separate brains, or if the mind does not come from the brain at all, but is something we download on a global scale at certain shift points in the timeline of evolution of our planet.

THE CALENDAR OF THE MAYA

The whole issue of what caused the scientific revolution comes in a new perspective when we realize that the year 1617, which historically speaking seems to have been its actual beginning, is a very significant shift point in the ancient Mayan calendar system. This year a so-called baktun of 394.3 years (144.000 days or 400 360-day years) began, which ended in the year 2011 (for an overview of the time line of the Long Count see Fig 3). Maybe then the current discussion about the relationship between mind and matter reflects that we are now again entering an era with a new thinking. In the same way as a new way of thinking and philosophy, that of the rational mind, was introduced at the previous baktun shift in 1617 another profound shift in how we understand reality may be emerging at the present time after 2011. At the very least, this is what would be consistent with the Mayan philosophy of time.

There is in fact massive evidence that all paradigm shifts in human thinking and especially the relationship between mind and matter are adequately described by shifts in the Mayan calendar. This has simply not become widespread knowledge in the academic community, which over all has refused to even study the matter. In the Mayan view, the relationship between mind and brain is not constant but subject to a wave movement, so that it undergoes significant changes at precisely defined points in time. As a corollary, the mind is not a product of our individual brains, but something that we "download", which shapes our perception and creation of the external world. This is not to say that the mind-brain relationship is simple, but the historic perspective based on the Mayan calendar provides an absolutely necessary inroad to understanding what the human mind is. This perspective tells us where the mind comes from, and more broadly what drives human history and the evolution of civilization. If the relationship between mind and brain is not static, but subject to a wave movement it seems obvious that we can never fully understand it merely from a physiological study of the brain. The key will instead be to understand the nature of the wave movement, which determines the evolution of the mind.

Overall, the global mind that we as individuals are all in resonance with is something that is activated during certain periods of time (certain baktuns of light, so-called DAYS) and deactivated during baktuns that are NIGHTS (or valleys in the wave movement. This mechanism is behind the rises and falls of civilizations that from a modern perspective often have appeared inexplicable or at least enigmatic. New civilizations and new technical abilities are invariably the results of the activation of a global mind

A critical aspect of this discussion is then how to define the mind. I believe that only if we use a strict and rigorous definition of the mind can we deal rationally with the issues at hand. The definition that I am proposing is that the mind is primarily a geometric structure.

The ancient Mayan inscriptions talk about a grid work of straight and perpendicular lines that was activated on earth in the year 3115 BCE as their so called Long Count calendar of thirteen baktuns began (see Fig 3). At this point in time, the human mind started to resonate with this global grid and it was through this resonance with straight lines that the individual human mind got its structure, and framework for understanding reality. Here, the mind thus does not refer to the stream of thoughts and visions, but rather to that which structures our thoughts and vision. The mind in this sense is a metaphysical structure that undergoes quantum shifts at predetermined ponts in time and leads us to look upon the world in new ways.

Similarly, in the Vedic tradition there was a significant shift at 3015 BCE when the so-called Kali Yuga began. This has later sometimes been misinterpreted as an astrological event and sometimes as related to the Goddess Kali, but it now seems clear that the way the ancient Vedic civilization looked upon this was that it meant the descent of the demon Kali, which was associated with division and quarrel. This of course can easily be connected to the separation and divisiveness, which is the very nature of the global mind. I take it for granted that the Hindus were referring to exactly the same event as the Maya did as a starting point for their Long Count. Yet, they were not as calendrically exact as the Maya were and so missed the starting point, or in other words the point in time at which the global mind grid could be downloaded, with about 100 years.

THE ORIGINAL RISE OF RATIONALITY AND CIVILIZATION

For the most part modern researchers are simply ignoring such information from ancient people with some kind of thought that they did not know better. In reality however the evidence is massive that human civilization began at the beginning of the Mayan Long Count (which is the same as the Hindu Kali Yuga), at a quantum shift in that activated the global mind.

First cities	(Sumer, 3200 BCE)
First pyramids	(Egypt and Sumer, 3100 BCE)
First monarchy	(Egypt, 3100 BCE)
First numerals	(Sumer, 3100, Egypt 3000 BCE)
First measures	(Indus Valley, 3250 BCE)
First writing	(Indus Valley, 3250, Sumer 3200, Egypt 3100 BCE)
First money	(Sumer, 3000 BCE)
First calendars	(Egypt, Sumer 3000 BCE)
First wheel	(current-day Slovenia, 3250 BCE)
First bronze	(Anatolia, Crete and Sumer, 3000 BCE)

Fig. 1 Calleman

Phenomena associated with civilization emerging around the beginning of the Mayan Long Count in 3115 BCE (sources Wikipedia).

Figure 1 for instance lists a number of the phenomena that we most associate with civilization and when according to professional archeologists these first appeared on our planet. What we may then notice is that they all appear at the beginning of the Mayan Long Count (or the Vedic Kali Yuga) ± 100 years. The dates occur within a very narrow range and their uncertainty may simply be connected with the inherent uncertainty in 14-carbon datings of samples of this age. Mainstream historians are however still ignoring the Mayan calendar as a timeline for human history. Would it not be about time that we take the ancient sources seriously and approach them in an unbiased way?

All of the advances listed in Figure 1 can also be seen as related to the introduction of the straight line, a line of separation that among other things separated mentally the left and right hemispheres of the brain. What in other words happened at the beginning of the Long

Count was that the human brains on a global scale started to download a metaphysical mind divided primarily into two hemispheres, the left and the right. The difference in functions between the left and right hemispheres of the brain hence do not go back to differences in the grey matter on the two brain halves, but to the downloading at this point in time of a mind, which separated the functions of the brain halves.

Without the downloading of a mind with straight lines and boundaries separating the functions of the brain humanity would never have been able to create a civilization and none of the advances listed above would have appeared. The reason civilizations emerged at this time was that the individual human minds were (and still are) part of a common global mind that started its evolution at the beginning of the Mayan Long Count. This global mind was indeed as the Mayan inscription says eight-partitioned, but included a primary yin-yang-duality between the left and right brain halves that was more fundamental than the other partitionings, (or divisions). Very prominent symbolisms for this (called the Tree of Knowledge of Good and Evil in the Hebrew Bible) exists from cultures from all over the world (see Fig 2) and yet it seems historians have not noticed that the straight line, and the divisions it create was a novelty that was absolutely crucial (literally speaking) for the rise of human civilization. There are no straight or perpendicular lines (2) in nature. They only exist as a result of the emergence and evolution of human civilizations.

Mexico   **Mesopotamia**   **China**

Fig. 2 Calleman

Symbols of the duality of the Tree of Knowledge of Good and Evil from different parts of the world.

The symbols in Fig 2 are then each in its own way, and each from a particular part of the world, a symbol of the global mind that began to be downloaded in 3115 BC. They show the dualist polarity filters that were created for the human perception through our resonance with the global mind grid at the time and they also show how straight lines then became a crucial factor determining human thinking. Even though there are differences between the symbols from cultures in different parts of the world, the similartities between them are even more striking, something that shows that the mind that was then downloaded all over the world was global in character.

Partitionings, Straight Lines and the Building of Pyramids

Naturally, if a global mind with straight and perpendicular lines was activated at the particular time when the Mayan Long Count began we would expect to see this manifested somehow in the technological accomplishments of people all over the Earth. This is also exactly what you find as shown in Fig 3. We can there see that at the beginning of the Long Count pyramid building began all over the planet, not only in Egypt and Mesopotamia but as far away as in Peru (Caral) and Mongolia. And before this point in time of downloading of a fully developed mind pyramids are not found anywhere (at least not if you go by what has been verified by professional archeologists). To me at least, the simplest and most logical explanation to this is that the human mind is global in nature and as it was activated in 3115 BCE it became something that people everywhere could download and manifest in their physical reality. Pyramids, which have tantalized the human imagination for a very long time, are then the first reflections of the straight and perpendicular line that were inherent parts of the global mind that people started to download at the beginning of the Long Count (see Fig 3). We may for instance notice that the Great Pyramid at Gizah actually is eight-sided similarly to the symbols in Figure 2 and might, just as well as the other pyramids in other parts of the world, have been built to a large extent in order to honor the new mind that became available then and created human civilization.

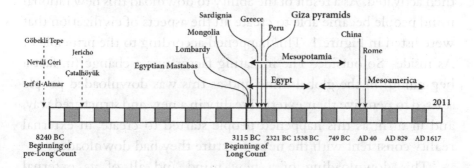

Fig. 3 Calleman

The building of perpendicular structures (broken lines) and pyramids (filled lines) in different parts of the world as related to the Mayan calendar and its preceding pre-Wave.

We should also notice in Figure 3 that even if pyramids only began to be built in 3115 BCE simple constructions with straight and perpendicular lines began to be built already at the beginning of the pre-Long Count in 8240 BCE. Hence, even if they did not come close to the precision of the later Egyptian pyramids, pre-historic sites such as Jericho and Catal Huyuk in current-day Turkey have houses that are rectilinear. In the Syrian site of Jerf- el Ahmar it is in fact possible to see how the first rectilinear structures emerged in a layer that has been dated close to the beginning of the pre-Long Count in 8240 BCE. This indicates that already at the beginning of the pre-Long Count an early form of straight and perpendicular lines became part of the human mind. This mental change prepared for the remarkable exact straight lines that came with the first pyramids at the beginning of the Long Count proper in 3115 BCE.

Based on the ancient inscriptions from the Maya and from India it thus seems that 5130 years ago a global mind was activated that included straight and perpendicular lines. These lines in turn created separation and discernment among the human beings and a whole range of mental faculties; writing, counting, creation of art etc, were effects of the integration of straight lines in the global mind that was then activated. As a result of the ability to download this new rational mind people became able to create all the aspects of civilization that were listed in Figure 1. This happened according to the principle of As inside- So outside. The initiating factor was a change (in fact a beginning) in the global mind and as this was downloaded people started to perceive their external reality in a new and structured way, and in turn, as this happened people started to create an external reality consistent with the new structure they had downloaded.

This downloading of a new mind and all of its external manifestations may have happened most obviously in the river cultures of the Nile, Euphrates, Tigris and Indus, because the natural conditions for agriculture and advanced civilizations were the most favorable there. But as we can see from Figure 3, the downloading of a new mind happened all over the planet so that large monuments with rectilinear structures were built in many different places, also where the continued life of a civilization could not be sustained.

The activation and deactivation of this rational global mind can then be followed as a wave movement continuing into the future as can be seen in Fig 3. Whenever a DAY begins in this wave movement the rational separating mind is activated generating cultures and civilizations consistent with this. NIGHTS however are different character. Their modes of thinking may be considered as more integrated (in contrast to analytical) but sometimes also more superstitious, the positive and negative aspects of the era we are now ourselves entering after the baktun shift in 2011.

The model presented allows us to understand why the modern scientific revolution began at the beginning of the seventh DAY of the Long Count starting in 1617. It also tells us that we have now entered an era of a NIGHT in the Long Count, the seventh NIGHT, which accounts for the shift in consciousness that is currently taking place. This, of course, is a simplified description of the history of civilization and the serious student of the global mind needs to consider many other factors that there is not room for to discuss here. In my view it is however not possible to understand the phenomenon of the mind outside the context of human history and how this relates to the Mayan calendar. As a corollary it is not possible to understand the phenomenon and origin of consciousness, not to mention the evolution of consciousness, without grasping the origin of the mind.

HOW THE GLOBAL MIND IS CREATED BY THE EARTH

Already from the notes above it should be obvious that the rational mind is not something that is created by the human brain. The human brain is believed to have been essentially the same throughout the 160,000 years of existence of *Homo sapiens*, and yet the human mind has undergone very dramatic changes in the same time period. Hence, we have to find the sources to the quantum leap of civilization about 5000 years ago in something different than brain changes and what I suggest is that it was caused by the activation of the global mind at that time.

As I have elaborated in detail in *The Global Mind and the Rise of Civilization* the global mind is something that is transmitted to the human beings by the Earth, and the polarity of the human mind shown in the symbols in Fig 2 are reflections of the four (or if we like eight in a compass rose) directions of the Earth. This connection between the civilized mind and the four directions may explain why the ancient peoples of our planet so much honored the four directions not only in ceremonies, but also in their designs of temples, monuments and cities. The Great Pyramid at Gizah for instance, differed from a perfect alignment with the northern direction (true north, not electromagnetic) with as little as 4 arcseconds. The reason is that it was the earthly gridwork of the four directions that by resonance of their brains, provided them with the structure of their mind and in a second step with civilization. The early civilizations of our planet were in fact happy to have been given civilizations from the gods and so wanted to show their gratitude to the four directions, which were real powers that had created their civilizations,

This would all mean that the human brain would need to be in resonance with the earth in order to be able to download a global mind. How could this proposition be demonstrated more directly? This question takes us to the enigma of the origin of the brain waves. For almost a hundred years it has been known that the human brain, or different parts of it, oscillates with low frequency electromagnetic waves, termed alpha, beta, delta and gamma, which are all associated with different mental states or states of consciousness. Science has however been unable to explain what generates these brain waves and their particular frequencies to begin with. What would be generating these particular frequencies? What would make brain cells fire in synchrony in widely different parts of the brain? No credible answers have been provided to these questions by the science of neurology. If we however recognize that the human mind is global in nature and that so the brain must be in resonance with the earth in order to be able to download the mind the enigma can, as we will see, be resolved.

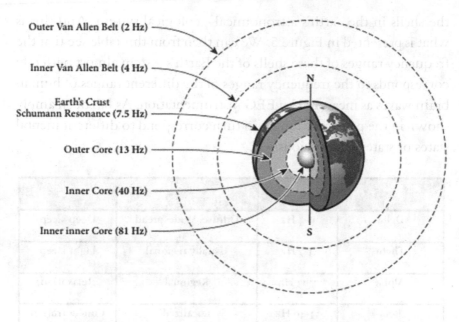

Outer Van Allen Belt (2 Hz)
Inner Van Allen Belt (4 Hz)
Earth's Crust
Schumann Resonance (7.5 Hz)
Outer Core (13 Hz)
Inner Core (40 Hz)
Inner inner Core (81 Hz)

Fig. 4 Calleman

Electromagnetic frequencies associated with different boundary spheres in the Earth's atmospheric-geological system.

In order to understand the origin of the brain waves we may use Figure 4 as the starting point. It has then been known for some time that the so-called Schumann Resonance at the earth's crust has a frequency, which is very close to the lower limit of the alpha waves of the human brain, typical of a meditative relaxed state of consciousness. Contrary to false rumors the Schumann Frequency is perfectly constant and cannot change over time as it is simply a result of the speed of light and the earth's circumference: 300,000 km/s / 40.000 km = 7.5 Hz.

The interesting thing is that we may calculate the frequencies not only of the surface of the earth, but also those corresponding to other boundaries between different shells in the earth's interior and its atmosphere (Fig 4) (Note however that the frequencies of the Van Allen Belts are averages since these belts are not spherical). What this means is that we can assign a frequency range to each of the shells in the earth's astronomical-geological system. And this is what is presented in Figure 5. We can then from this table see that the frequency ranges of these shells of the Earth's system almost perfectly corresponds to the frequency ranges of the different ranges of human brain waves as measured by EEG instrumentation. As has been amply shown in the past these ranges in turn correspond to different mental states or states of consciousness.

Brain Wave Type	Frequency	Distribution in Brain	Mental State
Delta	0-4 Hz	Diffuse, Widespread	Deep sleep
Theta	4-7 Hz	Usually regional	Light sleep
Alpha	7-13 Hz	Regional	Relaxation
Beta	13-40 Hz	Localized	Concentration
Gamma	40- Hz	Very localized	Higher states

Fig. 5 Calleman

The frequency ranges of different types of brain waves and the mental states these correspond to.

From this we may realize that the human brain waves ultimately depend on what shell in the atmospheric-geological system that the brain is in resonance with and that this resonance determines what state of mind that the human being is in. Our states of consciousness, or mental states, are thus not generated by our brains in isolation, but by what shell in the earth's system that our brains are in resonance with. This provides further evidence that the human brains download what in fact is a global mind from the Earth. It may also be argued that the reasons that there are different states of consciousness to begin with are A/ That the structure of the Earth is different at different resonance frequencies and B/ That the brain's ability to download such states may be subject to a number of physiological and biochemical factors.

To see an example of point A we may consider that when we are in sleep states our brains are in resonance with the atmospheric shells of the earth's system, where there is very little structure (as there is in dreams). If we go to the other extreme to the gamma waves those occur when our brains are in resonance with the earth's inner core, which because it is a crystal made up from iron and nickel, is the most structured part of the earth. Hence, when our minds are involved in the most advanced activities requiring structures, straight lines etc. then our brains are in resonance with this part of the earth and display EEG frequencies in the gamma range. Human civilization was then created by the activation of a global mind emanating from the structured inner core of the Earth, where straight lines have been inferred to occur.

From these observations it seems clear that the human mind is not something that is created inside our brains, but emerges from a resonance of these brains with the earth's system of different shells. We all in other words have part in an evolving global mind. Of course, the idea of a collective consciousness is one that has been around for some time, but what is new with this work is that we may understand how this collective consciousness emerges and how, in a wavelike manner its nature undergoes change over time. The delimitations of the evolution of the mind in space and time have been defined and we can now really understand the evolution of humanity as a function of a wave. The pendulum swings in history and civilizations rise and fall and yet, the evolution of consciousness has a direction, meaning that we are part of the manifestation of a higher purpose, even though we may not always be able to fthom this through the resonance of our brains. Expanded studies of the Mayan calendar in its totality of nine such waves allows us to understand not only the origins of our minds, but also where we as a species are going in a very broad sense. This will be the topic of Volume 3 of the Paradigm Shift Trilogy that I am currently writing.

We may then also immediately understand the phenomenon of synchronicities that was elaborated by Carl Jung or the common appearance of independent and simultaneous discoveries that have been made by different researchers. Already the building of pyramids in different parts of the world at the beginning of the Long Count in Figure 3 can be called a synchronicity, a synchronous independent creation of the same thing in widely different parts of the world that were not in contact with one another.

Throughout the course of history there have been many other cases like this. The independent and simultaneous development by Leibniz and Newton of calculus is one and the invention of the telephone by Alexander Graham Bell and Elijah Grey another. New creations do not occur at random points in time, but because of the predetermined wave movement of the global mind that people everywhere are in resonance with they will appear simultaneously in different parts of the world. Because of this wave movement, and only because of this wave movement, there is an inherent "timing" to everything that happens on this planet. There are things "whose time has come." We are all connected through our resonance with the Global Mind and this is why also on a more everyday scale we experience synchronicities. Everything is connected to everything else and we are living at a time when we are increasingly becoming aware of this ultimate unity of consciousness.

What is given here are only a few examples of phenomena that are the products of the global mind. In reality also religions, technology, art, governance, economy and a host of other things are following the evolution of the global mind. The very existence of such an evolutionary process is also what is behind the ancient idea that prophecy is possible. The global mind that we download should however probably best be likened to different versions of a software program, like Word 1.0. 2.0 etc. What this means is that even if we share a common structure for organizing our thoughts, those thoughts are not identical. A software program allows for a considerable variation in how it is used by different users, and yet to be effective it needs to provide a common framework. This framework also evolves as new versions of it are being introduced. The beginning of DAYS in Fig 3 may then be looked upon as more and more advanced versions of the global mind. Such beginnings and the corresponding upgrades of the global mind allow for a variety of new inventions and novelties of thought.

These kind of phenomena would never appear if the brains of each one of us would independently be creating an individual mind. If this were the case we would hardly ever be able to understand one another. The reason that we after all do understand each other so well is that we are operating by the same overall framework given by the global mind and its evolution. In *The Global Mind and the Rise of Civilization* hard evidence is presented supporting the idea that we all have part in one and the same global mind, which generates a collective consciousness. What may make the model presented in it different from other models is the fact that it presents the evolution of consciousness and a mechanism according to which this may occur through quantum shifts in the mind. Through this model we can see not only the indivisibility of our existence, but also that in certain eras we have come to perceive reality as very divisible. Concepts as the mind or consciousness are not static, but can only be meaningfully probed in an evolutionary perspective that allows us to understand that over time there are paradigm shifts and why our world views change.

REFERENCES

CJ Calleman, Solving the Greatest Mystery of Our Time: The Mayan Calendar, Garev, London UK 2001.

CJ Calleman, The Mayan Calendar and the Transformation of Consciousness, Bear and Co, Rochester VT 2004.

CJ Calleman, The Purposeful Universe, How Quantum Theory and Mayan Cosmology explain the Oriign and Evolution of Life, Bear and Co, Rochester, VT 2009.

CJ Calleman, The Global Mind and the Rise of Civilization, A Novel Theory of our Origins, Two Harbors Press, Minnepapolis, MN 2014. A second edition will be published by Bear and Co n 2016.

ALL IS ONE: THE ONE, THE UNIVERSE AND CONSCIOUSNESS

Dr. Attila Grandpierre
Konkoly Observatory

The ordered universe, which is the same for all,
was not made by one of the gods or by humans.
Rather, it always was, is now, and forever will be an ever-living fire,
ignited in measure, and extinguished in measure.

Heraclitus, Fragment 30

ABSTRACT. WE ARE IN search of the Absolute, the ultimate fix point of the world, the self-existing entity on which we can stand our footing firmly. To reach that aim we consider the relevant achivements of the last millenia in the light of the enlarged, postmodern science that preserves all the power of physics yet generalizes it to take into account the own laws of life and consciousness. In our search we are led to recognize the essentially complete view of the Universe consisting from three levels of reality: phenomena, laws and first principles. We have found three first principles of the Universe, that of matter, life and consciousness. We show that the architecture of the Universe has a fundamental triune structure corresponding to the first principles of physics, biology and consciousness. We point out that this fundamental Trinity of the Universe is the product of a still deeper, unmanifested entity. We found the ultimate, self-existing reality what we call as the One, the unified whole of the life principle, the Cosmic Self and the consciousness principle. The One is the realm of infinity, the eternal source of finite things. We show that although it is not self-existing, the existence of the All does not rely upon the existence of anything else. We show how the One permeates the All by the Order of Nature including the cosmic organizational principle, the principle of rationality and the Cosmic Self. We show that in the Order of Nature the One is the alpha of reality and the All is the omega. Surprisingly we found that the principle of non-contradiction, the axiom of logic can be derived from the ultimate nature of the One. We mention some related ideas of ancient myths like the Cosmic Egg and the World Tree, and philosophers like Heraclitus, Proclus, Spinoza, Christian Wolff and Whitehead. We indicate the signficance of the One in our personal life to live uncompromising life.

Keywords: science, laws of Nature, first principles, self, self-existing entity, substance, infinity, logic, axioms

THE PRIMARY FACT OF OUR LIFE

The primary fact of our life can be formulated shortly by the following short sentence: *we are living in the world*. The three players of the game are: (1) we as human beings, having consciousness to know life and the world, (2) life that mobilizes us to act in the world, and (3) the world in which we have to act.

We have to act for our life. We have to think how to act. To receive the results required by life from the world, we have to know the world. In scientific terms, we are conscious beings, and consciousness has the fundamental task to know life and the world, ultimately, life and the Universe. Consciousness has the natural task to obtain the relevant and reliable knowledge about the Universe. Considering that what we need is to foresee the changes of the world, what consciousness needs is the knowledge about the laws of changes, a useful picture of the Universe, a worldview including its most fundamental laws.

The road of consciousness starts from us and its purpose is to reach the Universe. Consciousness has the natural task to show the way towards the optimal symbiosis of our life with the world, ultimately, with the Universe. In the light of recent scientific results we may realize that life is present on Earth since four billion years and during this cosmically significant timespan it made all efforts to develop life and consciousness in an increasing rate of complexity and sophistication. Life on Earth manifests itself like a highly efficient and active cosmic creative power. Astrobiology tells us the life is a cosmic imperative (Davies 1999, 2006; Grandpierre 2017a). In order to live our life in accordance with its genuine nature, we are in search of the motivating power of the Universe that urges us to live our life in a meaningful way.

A SHORT DEFINITION OF THE BASIC CONCEPTS: MATTER, LIFE, CONSCIOUSNESS, UNIVERSE

We define an object as 'matter' when its behavior can be described by physical laws.

The best definition of life known to us is given by Ervin Bauer. *"The living and only the living systems are never in equilibrium; they unceasingly invest work on the debit of their free energy budget against that equilibration which should occur for the given the initial conditions of the system on the basis of the physico-chemical laws"* (Bauer, 1967, 51). This means that life has its own fundamental principle which makes the behavior of living organisms characteristically different from the behavior of physical objects.

Consciousness is defined as being the realm of thinking, including the laws of thinking and, at its 'center', the self selecting the goals. The self is defined, in scientific terms, as the ability to make decisions autonomously, that is, not completely determined by physical and biological factors. *The self is the bridge between the realm of thinking and the realm of matter,* because decision making is what initiates actions in the world (Grandpierre and Kafatos 2012, 2013). Consciousness is therefore fundamentally the unified system of the way we interpret life and the world. This world interpretation system is called shortly as worldview. We percieve the world and life with our external and internal senses, perceiving our percepts, feelings and thoughts. Each worldview has its own logic, fundamentally depending on the relative weight of the ontological categories. These perceptions serve as input data for our worldview-dependent logic that makes the decisions (Grandpierre 2014a). The self is the Key of the Universe because it is the only way in which the Universe can act in the finite, material mode of existence.

The Universe is defined as the unified whole of everything that exists. To determine what exists is the task of ontology. The first task of ontology is 'carving Nature at its joints', to find the most useful first divisions of the Universe as a whole. We found the most useful first division of the Universe the following: matter, life and consciousness, corresponding to the fundamental sciences of the Universe: physics, biology and psychology. Modern civilization is based on the idea that fundamentally only physical matter exists. This view is underpinned by the apparent fact that there exists only one exact science, physics. The development of an exact biology hallmarks a new, postmodern civilisation. In the followings, we present an overall picture of the Universe.

THE CENTRAL ROLE OF THE UNIVERSE IN OUR LIFE

Since ancient times man accommodates himself to the Universe as a whole (Couderc 1964). "Man does not exist for long without inventing a cosmology, because a cosmology can provide him a world-view which permeates and gives meaning to his every actions, practical and spiritual" (Kuhn 1995, 6). "Cosmology should not be confused with astronomy. The proper object of the study of cosmology is not the stars but the universe considered as a whole" (Heyneman 1993, xxiii).

The central role of the Universe in our life is a fundamental fact. Our life is produced by the laws and conditions of Nature. The Homo sapiens is a member of the biosphere that is the offspring of cosmic creative powers acting on our planet Earth, one of the planets of the Solar System arising from the Milky Way galaxy which is the offspring of the Universe. These cosmic creative powers are the ones creating life and consciousness. Our life and consciousness are the manifestations of cosmic life and consciousness. These cosmic creative powers connect us in the most intimate way to the Universe. We can realize ourself when we realize that the ultimate motives of our life are given by the Universe. Our personal life is a thread in the cosmic life. We are players in a cosmic game. Our terrestrial life is a mission on planet Earth. Our body is the spacesuit of our soul being a space traveller. We are cosmic ambassadors. Our self has a freedom over the matter of our body. This freedom over matter includes an ontological depth beyond the gigantic network of physical causes. Considering that the gigantic network of the physical causes extends to the entire physical universe, our self has a cosmic status. Our self is the outposts of the Cosmic Self.

MODERNITY AND THE FRAGMENTATION OF OUR WORLDVIEW

Since ancient Greeks, the view claiming that the only real existents are the physical objects accessible for our outer senses has become increasingly popular. Thales was in search of the first principles of the Universe and found it in water, that is, in a material entity. It was a frequent view in Newton's time that bodies have a complete, absolute, and independent existence (Westfall 1994). Nowadays the apparently dominant view among philosophers is that the Universe does not exist as a big cosmic whole. "The doctrine tells that all there is to the world is a vast mosaic of local matters of particular fact, just one little thing and then another" (see Lewis 1986, ix). In the modern worldview, the world as a whole is conceived as broken into material fragments, and what remains in this picture is a vast mosaic of finite, self-existing physical objects. In physical cosmological models the domain of material universe observable for us is a finite object.

We would like to point out here that this "little thing" picture is superficial and in a deep sense it is in direct conflict with the most significant achievements of physics. Physical laws play a central role in physics. Modern physics tells that all the material of the Universe has been created from quantum vacuum fluctuations by the physical laws (e.g. Tryon 1973, Hawking 1988, 142; Davies 1992, 73). This means that in this respect the physical laws precede matter. *The existence of matter assumes, fundamentally, the existence of physical laws.* Both the existence and durability of physical bodies are due to the invariable existence of physical laws. Our point here is that physical laws are *unifying laws*. They prevail everywhere in all the space. This fundamental fact calls attention to *the existence of physical laws*.

We have to distinguish between *physical equations* describing reality and *physical laws* that exist in reality. Physical equations exist in human mind but physical laws exist in Nature. Physical laws of Nature represent causal power (e.g. Armstrong, 1993, 422). *It is causal power that distinguish abstract from real entities.* Physical laws govern the interactions of physical objects (Roberts 2008). In the absence of physical laws, physical objects could not interact. Without interactions, physical objects could not exist at all. This means that the existence of laws of Nature is distinctively a robust fact. *The existence of observable matter is conditional since it depends on the existence of physical laws.* These physical laws themselves are unobservable.

MODERNITY AS THE ROAD TOWARDS POSTMODERN SCIENCE

The development of modern physics has paved the way towards the development of an exact science of life and consciousness. Within the conceptual framework of modernity, the origin of physical laws seems to be a scientifically unsolvable problem. As it is formulated recently, "it seems that almost all physicists who work on fundamental problems" accept that "the laws of physics stand at the base of a rational explanatory chain, in the same way that the axioms of Euclid stand at the base of the logical scheme we call geometry" (Davies 2004). We have to note that in natural sciences the things playing the same role than the axioms in mathematics are not the laws of physics but the first principles. The first principles of sciences can be regarded as the ultimate laws from which all the other fundamental laws can be derived. David Hilbert, one of the most outstanding mathematician of the 19th and the early 20th century postulated that the least action principle can play the role of the fundamental axiom in physics, from which all the fundamental physical laws are derivable (Hilbert 1918, 415/1115; Stöltzner 2003).

We arrive at two important achievements. One is the existence of physical laws which transcends the material mode of existence. The other is the fact that all the fundamental laws of physics can be derived from the least action principle. This means that the existence of physical laws is rooted in the existence of the least action principle. As the physicist Anthony Zee has formulated: "The action principle turns out to be universally applicable in physics. All physical theories established since Newton may be formulated in terms of an action. The action formulation is also elegantly concise. The reader should understand that the entire physical world is described by one single action." (Zee, 1986, 109). The least action principle is the source of all the fundamental physical laws.

The laws of physics has also only a conditional existence. The existence of physical laws depends on the existence of the physical principle. Ultimately, the existence of physical laws is rooted in the existence of the least action principle. In the same way as the material objects could not exist in the absence of the physical laws creating and governing them, the exsitence of physical laws should be regarded also as conditional because their existence and governance through their input conditions is given by the physical principle. In our search for the self-existent world, we have to consider in the next step the existence of the physical principle.

LAWS OF NATURE AND INITIAL CONDITIONS

In order to answer this question we have to consider first the role of initial conditions for the physical laws. Is there a room for anything else in the world beyond physical phenomena, laws and the physical principle? Let us call attention to the fact that *the equations of physics in themselves are not enough to describe concrete processes.* The fundamental equations of physics are differential equations describing only the changes of physical processes. *In order to obtain concrete process, initial conditions are necessary.* These initial conditions are crucial because they breathes fire into the equations, they are the ingredients which make physical existence possible. These initial conditions are the *input data* for the equations of physics.

The complexity of Nature can be conceived, with the mindset of theoretical physics, as present in the initial conditions, the input elements of physical laws. With the words of Eugene Wigner, the Hungarian–born Nobel laureate, one of the fathers of quantum theory: "The elements of the behavior which are not specified by the laws of nature are called initial conditions." (Wigner 1995, 322). This means that the complexity present in the initial conditons of physics cannot be determined physically.

Complexity is present in an extremely high rate in living organisms. In living organisms, the "input conditions for physical laws" themselves are highly organized in space as well as in time. This organization is called as *biological organization.* Biological organization is closely related to biological behavior, because structure and function are two aspects of biological behavior. This means that the input conditions of physical laws can be organized in a special way to fulfil biological functions, that is, biological aims. If biological behavior has its own, specific, biological laws underivable from that of physics, than new, biological laws of Nature exist beyond the extreme complexity of the input conditions of physical laws. If so, biology may be regarded as the control science of physics, considering that the input conditions of physical laws are regulated biologically for the sake of biological aims (Grandpierre 2007).

We have proposed (Grandpierre, Chopra and Kafatos 2014) that there is room for biological organization, namely, beyond the quantum level. This means that biological causes can continuously modify the physical conditions in every elementary, quantum time step. In every quantum time step the energy is undetermined within the limits of the Heisenberg uncertainty relation. Biological causes can modify the behavior of every elementary particle within these physically undetermined quantum limits. Remarkably, biological causes may modify physical conditions according to biological aims and elicit the physical causes necessary for it.

THEORETICAL BIOLOGY IMPROVING AND EXTENDING THEORETICAL PHYSICS

By our best knowledge, it was only one biologist, Ervin Bauer, who developed a systematic, consequent and adequately general scientific theory of life. Ervin Bauer has successfully determined the essence of life in exact scientific terms. He recognized that life has its own natural laws, and the essence of life can be found in the universal principle of life determining the behavior of all living organisms (see above). It is Bauer's principle that was helpful to make the next step, the generalization of the physical principle of least action. Bauer's theory was recently reformulated in a way that can be regarded as a fundamental extension of modern theoretical physics, generalized at its deepest level (Grandpierre 2007, 2011a,b; Grandpierre, Chopra and Kafatos 2014).

In our formulation, the first principle of biology is the generalization of the least action principle by a minimal but most powerful step. In physics, the endpoint of the least action principle is determined already by giving the initial conditions for it. In biology, living organisms can determine their endpoint. *Generalizing the first principle of physics by allowing the endpoint of the action principle to be a free variable and determined by biological aims* is the smoothest possible way to extend the action principle in biology (Grandpierre 2007). This move makes the most powerful tool of physics, the action principle, even more powerful. The so obtained theoretical biology is wider and deeper than modern theoretical physics.

If the biological principle is more general than the physical one, it must also permeate the entire Universe. This means we are not living in an inanimate aggregate of material universe, but in a biological, Living Universe (Grandpierre 2002, 2012, 2017). It is this move that makes it possible to extend the modern scientific worldview into a broader, postmodern worldview. This achievement has not only a scientific, but a philosophical significance. In a biological Universe life is not constrained to small occasional volumes within skins and membranes. Instead, life populates the entire Cosmos in an unexpected variety of different cosmic life forms (Grandpierre 2008), all the space between elementary particles and stars. Taking into account that the quantum-vacuum itself is a cosmic living organism (ibid.), we may realize that life exists unconditionally. The living nature of the cosmic vacuum exists unconditionally, because it is based on and vitalized by the cosmic life principle transcending physical space and time. The Bauer's principle tells that living organisms are living when they mobilize all their available energies towards the heights of life. Instead of inert matter, it is activity that is the most basic property of the Universe, activity for the sake of life. The Universe as a whole is a living organism (Grandpierre 2002, 2004, 2012, 2017).

If life has its own first principle, than life is not doomed to play a subservient role in a physical universe dominated by physical laws. If life has its own first principle and it can realize it on a cosmic scale, than it has its own nature, its own life, its own aim and own laws to follow and realize, independently of physical conditions and, ultimately, on physical laws. This makes sense since it tells that physical matter is the limiting case of life in the case of such living organisms that manifest practically zero ability to act and move their macroscopic center of mass in space. The physical principle is a special case of the biological principle. *It is not matter but life is the more fundamental reality.*

Living organisms manifest ongoing structural transformations that are elicited in a way transforming all the chemical energies into the structural energies of the cells. In this way, the physical conditions determining the biological processes are continuously modified in a special, biologically regulated manner necessary to modify the physical processes in a way that they lead, within these continuously and suitably varying conditions, to increase and regenerate biological potentials (Bauer 1967, 45, 49, 62). *This means that within living organisms biology organizes the input conditions of physical laws in a way that the output of physical laws be suitable for biological purposes.* It is the biological principle that is capable to breath fire into the physical laws.

Let us illustrate it by the following example. When we bend our finger, we have to make only one decision. All the rest goes automatically. The result of our decision is, when we think about it more closely, immense and deeply impressive, considering that an immense number of electrons, atoms and molecules will immediately change their motion in an extremely finely orchestrated manner just in one of the possible ways, the special one with suitably changing coordinated velocities that is necessary that their motions when seen from a higher level result the bending of the finger. All these immensely complex and sophisticated orchestrated motions occur by an extreme ease, as if these electrons, atoms and molecules were not physically inert objects changing their state of motion only when suitable strong physical forces act upon them.

All the necessary investment of physical work occur perfectly in the absence of our conscious care and calculation. Nevertheless, if a physicist should solve the task, he would need an immense amount of data about the initial positions and velocities of all atoms and molecules. Moreover, the physicist would need a suitable idea involving the basic connection between the physical data and the required bending motion of our finger. The physicist would need a suitable model of the biological organism containing the relation between decision making and the physical state of our body. Considering that the decision making works in case of any living organism capable to move their extremities, independently from their actual physical states, the task seems to be definitely unsolvable. In such biological problems, the so powerful method of physical approach proves to be inefficient and insufficient.

But once we realize that the relation between our decision already made, all the subsequent sum of physico-chemical changes is related to the first principle of physics, because it is the physical principle that determines the most efficient pathways in between the already determined endpoint and initial point, we obtain an enormously powerful insight into the physics of the process. The least action principle is ideally suitable for biology because it works in the most economic manner. This is why our body works by extreme ease, seemingly in an effortless manner, when we bend our finger, all the details are arranged without our conscious effort. *The decision making determines the energy landscape in a way that in the biologically determined energy landscape the minimal energy pathway leads from the initial state to the biologically determined endstate in the same way as, in a geographical landscape, a river flows from its source to the ocean. Every detail is arranged by biological determinations.* These biological determinations occur at the subquantum level, that is, at a level of Nature where physical conditions and laws do not prevail. This means that at the subquantum level biological determinations act freely. The body of living organism is a living landscape governed by life (Grandpierre 2007).

We had seen that the laws of Nature arise from the first principles. Now we can see that the first principles are suitable to govern the input conditions of the laws of Nature in a way similar to what the laws of Nature do, organizing the relations between observable phenomena. This means that both the laws of Nature and their input conditions arise from the first principles, with the assistance of biological autonomy.

ENDPOINT SELECTION AND THE FUNDAMENTAL PRINCIPLE OF CONSCIOUSNESS

We can term the activity exploring, interpreting, evaluating and selecting the possibilities suitable to serve as endpoints for the biological principle as conscious activity. We found that the natural task of consciousness is to explore and determine the best conditions for life. Exploring and understanding life and the world is not a small task. Consciousness must be capable to fulfil its natural task. Consciousness must be as deep and broad as life and the Universe are in order to assist life efficiently. Indeed, our biological theory tells that decision making ability is inseparable from life. If decision making is made by consciousness, this means that consciousness is inseparable from life. Considering that life is a cosmic reality, consciousness, its partner and servant, must be also a cosmic reality. If so, *consciousness is capable to understand the nature of life and the Universe.*

We arrived to the basic problem of consciousness. On what grounds may the biofriendly decision-making capability work? Could it work without a factor that ensures the conditions of correct decision-making in respect both of life and the world? Did Nature supply the factor necessary to correct decision-making in any situation? The nature of life demands a consciousness with reasoning capability suitable to evaluate the important inner and outer conditions, relations, individual and communal viewpoints properly. In order that decision-making could be correct, a guiding factor is necessary which is capable to follow the requirements arising from both the principle of life and of matter. This means that *a universal, reliable guiding principle is attached to the biological principle: the rational principle.* In other words, not only matter and life, but consciousness must also have its own fundamental principle. The rationality principle must be suitable to prescribe within any given condition the laws of reliable conjectures in a way that fits both at the physical and biological level. This means that the two basic properties of the rationality principle is the capability to guide consciousness according to the physical and the biological principle. Consciousness is the not the late result of cosmic evolution but one of its fundamental ingredients.

THE CAUSAL RELATION OF THE MODES OF EXISTENCE

Causality is a fundamental tool of cosmic activity. We found three fundamental types of causes: physical, biological and psychological ones, initiated by matter, soul and consciousness. The causal relations between them are enlightening. The biological and psychological causes are able to initiate and govern physical causes. As in the case of bending our finger, *it is the meaning of our decision to bend our finger in a specially organized manner suitable to elicit all the innumerable physical causes realizing the biological or psychological decision.* Although physical causes also can elicit biological and psychological causes, as in the case of a physical injury, the usual way of life is characterized by the priority of biological and psychological causes. Since the natural task of consciousness is to select the best endpoint for the biological principle in the given context regarding the lifespan, the priority among causes is given by Nature to life.

If the natural task of consciousness is to assist life in the best way possible, than there is a *moral world order* indicated by the principle to respect life and act for the best of it.

THE FUNDAMENTAL TRINITY

We found three first principles as the fundamental principles of the Universe: the principles of matter, life and consciousness. These three principles permeate and govern the entire Universe, all material objects, living beings and their consciousness. Among these three principles the principle of matter is a special, limiting case of the biological principle. Considering that the natural task of consciousness is to determine the most suitable endpoints for the biological principle, we found that the rationality principle is inseparable but subservient to the biological principle. We arrived to realize *the cosmic priority of the biological principle that includes the material principle and co-acts with the consciousness principle.*

The physical principle has a remarkable property: economy. The least action principle selects the most economical path for physical behavior in terms of action. The biological principle has also a remarkable property: it is a maximum principle directed to create the maximally favorable conditions for the development of life. Similarly, the consciousness principle has also a remarkable property. It prefers the most clear, elegant, simple and beautiful ways of thinking. The three fundamental principles can be formulated in a unified manner. Together they represent the three fundamental types of causality: physical, emotional and intellectual, in the most economic, maximally life-friendly, beautiful and elegant manner.

These three types of causality cover all the fundamental types of causes. The three fundamental principles of the Universe, together with the laws and phenomena inseparable from them, form one unified whole: the Universe. The Universe is unified by its fundamental principles. Similarly to laws of Nature, the first principles transcend the 3+1 dimensions of space and time, the material modes of existence. The laws and first principles of Nature exist in still more fundamental layers of reality in their own modes of existence.

We found three fundamental ontological domains, that of matter, life and consciousness. All together they form a *fundamental Trinity*, the cosmic Trinity of matter, life and consciousness. Remarkably, one of the most important symbols of ancient myths are the World Tree, frequently named also as the Tree of Life or Tree of Knowledge. This fits nicely with the cosmic Trinity we found here: the Trinity of the material universe, the biological universe and cosmic consciousness. These latter ones are also frequently referred to as the World Soul and Cosmic Mind, together as God. In panentheism, God is meant in broader terms including also the material universe.

DISCOVERING THE ABSOLUTE, ULTIMATE REALITY: THE ONE

We are in search of the Absolute, self-existent reality. Scrutinizing the first principles from this point of view, we have three important things to note here. The first is that the physical principle can be derived from the biological one as its limiting case. This means that the physical principle, although fundamental, is not ultimate itself. The second is that the biological principle is inseparable from endpoint selection, therefore from the self, therefore in the cosmic context from the Cosmic Self. The third thing to note is that the Cosmic Self is inseparable from its own first principle, the principle of consciousness. This means that the ultimate principle, the biological principle, is inseparable from the Cosmic Self and the consciousness principle. We can draw an important conclusion. The ultimate reality consists from the biological principle, the Cosmic Self and the consciousness principle. The biological principle, the Cosmic Self and the consciousness principle form an inseparable unit, the ultimate unit of reality.

The primary empirical fact of our life is that we have to act in the world. This possible because the matter of our body is merely a tool in the hands of life and consciousness, because the physical principle is a special case of the biological principle in the limit when the capability to act shrinks to practically zero. In the cosmic context, life and consciousness rules the All, the Universe as a whole. In the light of all the empirical and theoretical evidences, we argue that it is necessary that besides the observable physical universe a different, non-material Universe exists which is free from any material limitations but full with unlimited, supreme life and reason. We call this supreme, non-material but living and conscious Universe as the One.

It is an unsolved problem whether the non-material realities like the laws and first principles of Nature or the self may exist independently of matter or not. Empirical observations tell that our feelings and thoughts can live their own life largely or practically completely independent from the body. In a variety of mental states which are practically independent from our bodily life we experience inner processes as real life: *imagination, thinking, dreams, or out-of-body experiences*. Moreover, *life has a variety of matter-independent characteristics* (Popa 2004) *as well as information that it possible to become manifested in a completely different material forms: minds, books, tapes, movies etc.* Our theory gives a profound theoretical basis to these empirical evidences.

Our postmodern science leads us to recognize that beyond the fundamental Trinity of the Universe, there exist another Trinity, the ultimate Trinity of the life principle, the Cosmic Self and the consciousness principle: the One. We call attention to the fact that it belongs to the very nature of life to generate, maintain and propagate life. This means that life is a reality that exists by itself. Life as a cosmic reality is a self-existing, self-sustaining, unconditional reality: the life of the Absolute, the One. Life is the par excellence self-existing reality because it has a fundamental self-recreating nature. We can see with our own eyes and experience in our own life that living entities behave in that manner. Life is the active self-existing reality in eternal process. In this light our finding that cosmic life is inseparable from the Cosmic Self and the principle of consciousness means that we found the ultimate reality what have looked after. It is the Trinity of the life principle, the Cosmic Self and the consciousness principle. This unity is the single ultimate reality: the One.

All these discoveries shed light to the substantially deeper and wider realities beyond the overly narrow limitations of the modern scientific worldview. The Universe stands up before us as having a towering architecture with three levels of realities: that of phenomena, laws of Nature and first principles. The Universe has three ontological domains, that of matter, life and consciousness, all of them having their own modes of existence. We may notice the remarkable twofold triune structure of the Universe. Considering the immaterial or spiritual nature of the laws and first principles, we can term this remarkable structure as the fundamental intellectual architecture of the Universe. Considering that the most fundamental aspect of the Universe corresponds to the life principle, we can call the architecture of the Universe as biological. We have arrived to a scientifically guided organic cosmology.

We found that the living Universe, the All, the unified whole of everything that exist contains in itself another unified whole, the unmanifested Universe, the One. Beyond the Universe we are living in, there exists secretful ultimate reality that lives within us, too: the One. Our innermost life survives us as finite beings because our life at its innermost essence is eternal. It is the One. Keeping in mind the ancient idea of Mother Nature, we may express the same in a perhaps more suitable manner: She is the One. The Infinite Living One. Infinite Life in one person embracing Supreme Reason. The Absolute, self-existing reality living in us, most personally, at the ultimate level of our inner world. At the ultimate level, everything is one. Ultimately, All is One. We are One.

THE ONE, THE ALPHA AND THE OMEGA

We arrived to the end of the world. The One is the starting point from the viewpoint of Nature, and the end of our cognitive process. This means that the One is first principles are the alpha and the omega. The One is the alpha, because all the world of laws and phenomena starts from here. The One is the omega because the One is the ultimate object, the highest aim of human cognition. The One is infinite since the first principles of life and mind as well as cosmic biological autonomy are not finite; they are transcending space and time. The One is the Infinite One. The One is living, because its primary ingredient is the life principle. The One has a triune structure because it consists of three fundamental, inseparable ingredients: the life principle, cosmic consciousness, and the consciousness principle. This ultimate triune living organism we can refer to as the ultimate Trinity. The One is an infinite living being.

Life is by its very nature creative. Creativity requires life to be infinite in its ultimate nature because infinity can produce infinity plus finite world because, in the language of mathematics, infinite plus finite is equal with infinite. This means that infinity, life and creativity are one and the same reality from different aspects.

Since life itself is immaterial, it is capable to generate and govern matter with the help of the creative fundamental principle and the cosmically available free energies of the quantum vacuum. *Matter is governable by life and reason.* There exist biological and psychological energies by which we mean energies governed by biological and psychological causes. The will is definitely capable to mobilize real energies to generate biologically and psychologically determined outputs (Baumeister 2012). The biologically and rationally governed energies are capable to produce suitable virtual particle pairs for governing matter (Grandpierre 2012, Grandpierre and Kafatos 2012, 2013).

Ancient myths tell us about the shining cosmic egg that was the source of the universe. In vedic literature Hiraṇyagarbha, the golden egg is the source of the manifested cosmos. An ancient Orphic hymn addresses him thus: "Ineffable, hidden, brilliant scion, whose motion is whirring, you scattered the dark mist that lay before your eyes and, flapping your wings, you whirled about, and through this world you brought pure light."

THE LIFE OF THE ONE - LIFE WITHIN THE ONE

There is life within the One, because eternal feelings and thoughts can be experienced by the One, the Ultimate Being. Eternal ideas like that of mathematics, eternal feelimgs like that of music can be created in an innumerable number and this infinite set of eternal ideas and feelings can be experienced in an infinite number of contexts. This is the world of the Living One, in other words, Eternity, the realm of feelings and thoughts having eternal values.

It is a remarkable fact that we are free to wish what we like. There are no limits to our wishes. It is exciting to realize that we are free to think what we like. There are no limits to our thoughts beyond the internalized obstacles of the mind. We are, in reality, gods, gods of our inner world, of this inexhaustible inner universe. Our wishes and thoughts does not have to arise from empirical reality. Our feelings and thoughts may arise first of all from our nature-given thirst for truth, thirst of beauty, thirst of an intact, self-existing world of feelings and thoughts. Our thoughts and feelings may arise from our love of life and live life fully. The One lives its own life independently of any particular occasions. As Alfred North Whitehead has noted, "eternal objects" are the ones comprehensible without reference to some one particular occasion of experience.

THE LOGIC OF NATURE IS FUNDAMENTALLY IDENTICAL WITH HUMAN LOGIC

In actual reality, the road towards the appearance of the Homo sapiens starts from the Universe as a whole. The main stages of this road are the Milky Way, Solar System, Earth, the terrestrial biosphere and humanity. Human cognition walks the same way in the other direction, from the Homo sapiens to the Universe as a whole.

Heraclitus wrote about the logos as the Order of Nature. It is interpreted as the organization perceptible in things and the rational expression of it in words (Kirk 1975, 40). As Giordano Bruno has written about Nature's ladder: "Nature descends to the production of things, and intellect ascends to the knowledge of them, by one and the same ladder." (Bruno 1976, 93). Human intellect walks the road of cognition, Nature walks the road of cosmic evolution, the creation of the world and its changes. Considering that Nature's ladder, the road of cognition and of creation is one and the same from two different points of view, that of humans and the Universe as a whole, we can term this road with the name "the world bridge".

The identity of the road of cognition on which human logic walks and the road of creation on which Nature walks means that human logic and the logic of Nature are fundamentally identical. They should because we are within the womb of Nature and the fundamental principles of the Universe permeate us in the same way as they permeate the entire Universe. This means human logic is much more than formal logic. Nature's logic and truth-following human logic both are the same aspects the creative activity of the

One. *The All and in it and the world bridge from the One to the All are the products of the One with the help of Nature's logic.*

THE ONE AND THE ALL

Realizing that the ultimate unit of existence is the One, we have to distinguish it from the All. The All involves, besides the One, also the principle of matter, and with it the realms of finite entities, i.e. material objects, and their related aspects in the realms of life and mind. The All involves physical, biological and psychological phenomena, laws of Nature and the first principles, besides the unaccountably large number of biological autonomies, individual and collective consciousness. The All can be conceived as the eternal One plus the worlds of finiteness. The All is the Universe as a whole in process, including all the already happened phenomena, the laws and first principles and all the biological autonomies of the innumerable living organisms and their communities.

It may seem that the One and the All as two fundamentally different entities. The One is infinite, immaterial, eternal, unified and indivisible whole. In contrast, the All seems to be the sum of finite entities, each related to material entities, transient, temporary and spatial. Considering that the material principle can be derived from the life principle, one can consider the All as being ultimately identical with the One.

The All is potentially present in the One. The One is present in the All. The One is there at the depths of each and every entity endowed by finite material aspects. The One is present hidden in the All and lives its eternal life within all of us. The One is the ultimate basis and science of our personal and communal life. Together with all living beings, we live in a deep, intimate union with the supreme, inner Universe, the living infinite One.

We found that there are only one self-existing entity in the world: the One, that is, the unified whole of the life principle, the consciousness principle and the Cosmic Self. We found that there exist another kind of reality which as a whole does not rely upon anything else: the All. *The All is not self-existent because it is created by the One.* Nevertheless, because it contains within itself the One, *the All does not rely on its existence on anything else existing outside of it.*

The One and the All both live in a mode of existence which differs from the finite mode of existence. The All is the unified whole of everything that exists, therefore nothing exists besides it. The All does not have a boundary, in contrast to its material parts that have boundaries or spatio-temporal limits in the usual 3 + 1 dimensions of space and time. This means that the All, the Universe as a whole, is not material because it does not have material limitations. The All, the Universe as whole is not the direct object of our external senses. Instead, it is the reality beyond our mental, logically and scientifically constructed idea or concept trying to fit reality.

In the One cosmic feelings and cosmic thoughts are living their life. Cosmic feelings and thoughts are the realm of music and mathematics, aesthetics and ethics. The All is something more than the One since it includes finite things, the particular, occasional objects and processes the constitute the observable universe and the transient feelings and thoughts related to them.

THE FUNDAMENTAL AXIOM OF LOGIC ARISES DIRECTLY FROM THE ONE

The fact that the Universe as a unified whole is ultimately One is the most fundamental fact of the Universe, a fact that cannot be any other way. The Universe cannot be ultimately two or three because its fundamental principles are all-embracing, therefore unifying principles. Due to the unifying nature of the first principles, the All is necessarily one at its ultimate level. *The mathematical property of the One, namely, that its number is one, draws with itself the inevitable consequence that all existent entities should be consistent with each other.* The principle of non-contradiction is the direct consequence of the fact that the number of the ultimate entities is one.

The cosmic rationality principle can be conceived as the basis of Logic of Nature. We point out that the universal laws of logic can be derived from the One. *If everything is, ultimately, at the level of first principles, one unified whole, and if everything else arise from this ultimate unified whole, than in Nature's Logic nothing can contradict to anything else.* This means that the Logic of Nature has a fundamental property that can be formulated as the principle of non-contradiction.

Human beings have an access to the Logic of Nature. We can call the full form of logic accessible for human mind as the humanly experienced form of the Logic of Nature. Formal logic is only a part of this natural logic. Since logic is based on the principle of consciousness which is the ideal tool of the life principle, therefore natural logic has a primary commitment to proceed towards the fullness and heights of life.

Considering that the All is a unified whole that is at its ultimate level identical with the One, every actual process and material parts constituting the All should be consistent with each other. Contradiction between actual phenomena would be inconsistent. Therefore one of the ultimate property of the Logic of Nature must be the law of non-contradiction. Remarkably, the law of non-contradiction is the fundamental axiom of human logic.

The principle of non-contradiction is one of the fundamental axioms of human logic. All the parts of the All arise from the Whole according to the Logic of Nature, therefore should be consistent with the Logic of Nature. Of course, the constituents of the One, namely, the life principle, the reason principle and the Cosmic Self also should be consistent with each other, because they form a unified whole, the One.

Our result fits nicely with the unexplained facts of the unreasonable effectiveness of mathematics in science.

THE ONE AND THE FIRST PRINCIPLES - THE MAIN ROAD OF HUMAN KNOWLEDGE

At the origin of modern science Thales and the Greek philosophers initiated a break with traditions preferring a purified, restricted, more materialistic idea of science which qualified the wisdom of the ancients as myth. We think it is timely to recognize that ancient myths has preserved a wider and deeper knowledge about the Cosmos.

According to ancient, traditional knowledge, the term cosmic law designates the principle or set of principles believed to represent the most generalized nature of the order of things in the Universe. As the entry "Cosmic law" of the Encyclopedia of Religions writes: "Confidence in the existence of a principle of order in the Universe at large is, in turn, reflected in a common belief that individual events within society are not random (hence meaningless) occurrences but parts of larger meaningful patterns that extend through time. Furthermore it is this confidence – that the entire universe is established upon and governed by a principle of natural and moral order – that enables human beings, individually and collectively, to deal effectively with intellectual, moral, and spiritual life crises" (Long 1995).

Some of the earliest evidences about mythology tell that the idea of the One as the Supreme Being has preceded the idea of God. As the Rig Veda writes:

"The only One breathed breathless in itself,
Other than it there nothing since has been.
...
Then first came Love upon it, the new spring
Of mind - yea, poets in their hearts discerned,
Pondering, this bond between created things
And uncreated. Comes this spark from earth,
Piercing and all-pervading, or from heaven?
Nature below, and Power and Will above.
Who knows the secret? who proclaimed it here,

Whence, whence this manifold creation sprang? –
The gods themselves came later into being. –
Who knows from whence this great creation sprang?"
(translation by Max Müller 1859, 564)

In Chinese philosophy, 'Qi' or 'ch'i' is the vital principle of which everything is composed, "the ultimate foundation for the existence of the universe"...the "ontological source of the universe" (Zhenyu Zeng 2011).

In ancient Egypt, Amun is the One, Great and Hidden God of the Universe. The name "Amun" ("imnw") suggests imperceptibility in and of itself and derives from the verb "imn", meaning both "to conceal" and "be hidden" (Dungen 2017).

In the ancient Hungarian mythology "the unnamed Supreme Being is the One, preceding and surmounting everything else" (Ipolyi 1987, 4).

In the ancient Babylonian religious system the Supreme God, the first unique principle from which all the other gods took their origin, was Ilu, the One and the Good (Lenormant 1999, 113).

One of the last great philosophers of antique Greece, Proclus believed the true philosopher should become "a priest of the entire universe." In contrast to the skeptic position that the material universe is outside the human consciousness and can only be known through sensory impressions, the Neoplatonists emphasized the underlying unity of all things and considered that the first principle of the Universe is the ultimate One (to Hen). Neoplatonists thought of the One as the source of the good, or perfection, of everything. The One, including the divine Intellect, is beyond being, the "final cause" of all things.

We find remarkable similarities with the philosophy of Spinoza. By his Ethics, the observable nature is the product of "natura naturans" what is in itself, an eternal and infinite essence, that is, God.

As Ralph Cudworth, the leader of the Cambridge Platonists wrote in his book *The True Intellectual System of the Universe* (1678), according to ancient views the Universe forms a triune structure of ultimate principles. The first ultimate principle "delightful love, and that which is not blind, but full of wisdom and counsel", the second ultimate principle is the "infinite knowledge and wisdom" (Cudworth 1820, Vol. 1, p.426), while the third and last one is "infinite active and perceptive power" (ibid.).

Christian Wolff, the eminent German philosopher redefined philosophy as the comprehensive science of human knowledge founded upon the fundamental principle of logic, the axiom of non-contradiction, the very first principle of "all metaphysical first principles". His work relies on the idea that there is an intelligible order and interconnectedness between all the different disciplines.

Alfred North Whitehead, the founder of process philosophy has also noted that the order of Nature is given by general principles. "Geniuses such as Aristotle, or Archimedes, or Roger Bacon, must have been endowed with the full scientific mentality, which instinctively holds that all things great and small are conceivable as exemplifications of general principles which reign throughout the natural order" (Whitehead 1925, 5).

The idea of the intellectual structure of the Universe is present in modern science as well. The great discovery of the least action principle as the origin of all fundamental laws of physics has "imbued the philosophers of those days with an unbounded confidence in the fundamentally intellectual structure of the world." (Cornelius Lanczos 1952, xxii-xxiv). "After a lifetime of crabwise thinking, I have gradually become aware of the towering intellectual structure of the world" (Hoyle 1994, 423). In our work above, we found the intellectual architecture of the Universe as the "towering intellectual structure" of principles, laws and phenomena.

Modern science has obtained its most fundamental idea of comprehensive principles from the ancient tradition. As the Encyclopedia Britannica has formulated, the goal of physics "is the formulation of comprehensive principles, or laws of physics, that summarize disparate phenomena in the most general possible way". Indeed, since natural science should grasp Nature and Nature is all-comprehensive, it requires to find the most general, all-comprehensive principles.

A KIND OF SHORT SUMMARY

From time to time, the growth of science leads to fundamental achievements making a step closer to the original aim of science: obtain as complete and deeply penetrating knowledge about nature as possible. Here we reported on a fundamental improvement of the moderns cientific worldview. Our picture leads to the conclusion that ultimately we are cosmic flames, the flames of the One. Our consciousness is, ultimately, the light of the eternal cosmic flame. Our soul is, ultimately, the heat of the cosmic flame.

Humanity is driven by the deep and invincible conviction that our personal life is not finite, that true and infinite freedom must exist in our world, that the truth and goodness are the winners not only on the heavenly realms but this must be so on the Earth, too. In reality, the Earth is a celestial body, a planet of the Solar System.

These invincible convictions have a real and unshakable basis, because all these are rooted in the life of the One within us at the ultimate level of our inner world. We have to admit that the ultimate meaning of our personal present life on Earth is related to the ultimate meaning of all life in the Universe and is based on the actual presence of the infinite world process at the ultimate level of our inner world. Everything is realized within us because cosmic unifying principles are living in our inner world embracing the entire Universe. Everything is living and has its inner world. Everyone of us are endowed with the treasure of the One, the world which is infinitely rich in true, beautiful and good feelings and thoughts having en eternal value. Through the world of the living One within us, the realm of eternal truth, eternal beauty, eternal goodness is continuously urges us to act as fully living, integral and intact beings towards living the uncompromising life. At the ultimate level, life is the Absolute. Living with the power of Absolute, opening our mind towards the infinite cosmic principles driving our soul and mind we can live our life accordingly to our own, Nature-given way.

We are living in the world. Reason is living within us. Reason connects Man with the Universe. The Universe is the home of Supreme Reason and the Eternal Soul.

In our search for the Absolute, we have found it in the form of the supreme, non-manifested, living and conscious inner Universe. This achievement is due to the developments of philosophy, science and religion. In our days the time is ripe to find answers to ultimate questions. Humanity did developed a deep and wideranging knowledge that has become mature in our days. The development of postmodern science may offer a powerful tool to answer such metaphysical questions on the basis of its wider and deeper approach.

REFERENCES

Armstrong, D. 1993, 'The Identification Problem and the Inference Problem', Philosophy and Phenomenological Research, 53: 421-422.

Bauer, E. 1935/1967, Theoretical Biology (1935: in Russian; 1967: in Hungarian) Akadémiai Kiadó, Budapest, 51.

Baumeister, R. F. 2012, Self-control - the moral muscle. Psychologist 25, 112-115

Bruno, Giordano and Lindsay, J. 1976, Fifth Dialogue. Cause, Principle, and Unity: Five Dialogues, 93.

Cudworth. R. 1678, The True Intellectual System of the Universe. 1820, Vol. 1, 426.

Davies, P. 1992, The Mind of God. The Scientific Basis for a rational World. Touchstone, New York, 73.

Davies, P. 1999, The Fifth Miracle. The Search for the Origin and Meaning of Life. Penguin Books: London.

Davies, P. 2004, When Time Began, New Scientist Supplement, 09 October, 4.

Davies, P. 2006, The Goldilock Enigma. Allen Lane.

Dungen, Wim van den 2017, Amun, the Great God: Hidden, One and Millions. http://www.maat.sofiatopia.org/amun.htm

Grandpierre, A. 2002. Az élő Világegyetem könyve (The Book of the Living Universe, in Hungarian.) Válasz Könyvkiadó.

Grandpierre, A. 2004, Conceptual Steps Towards Exploring the Fundamental Lifelike Nature of the Sun, Interdisciplinary Description of Complex Systems 2: 12-28.

Grandpierre, A. 2007, Biological Extension of the Action Principle: Endpoint Determination Beyond the Quantum Level and the Ultimate Physical Roots of Consciousness, Neuroquantology 5: 346-362.

Grandpierre, A. 2008, Cosmic Life Forms. In: From Fossils to Astrobiology. Records of Life on Earth and the Search for Extraterrestrial Biosignatures, Seckbach, Joseph and Walsh, Maud (eds.) Springer: New York, 369-385.

Grandpierre, A. 2011a, The Biological Principle of Natural Sciences and the Logos of Life of Natural Philosophy: a Comparison and the Perspectives of Unifying the Science and Philosophy of Life. Analecta Husserliana 110, Phenomenology/Ontopoiesis Retrieving Geo-Cosmic Horizons of Antiquity, 711-727. Springer

Grandpierre, A. 2011b, On the first principle of biology and the foundation of the universal science. In: Astronomy and Civilization in the New Enlightenment, eds. Tymieniecka, A.-T. and Grandpierre, A. Analecta Husserliana, Springer, 107: 19-36, Springer.

Grandpierre, A. 2012a. Az élő Világegyetem könyve – 2012 (The Book of the Living Universe. 2nd, thoroughly rewritten version, in Hungarian.) Titokfejtő Könyvkiadó.

Grandpierre, A. 2014a, A Model-Independent Method to Analyze The Logic of World Models. Rome, January 2011. Phenomenological Paths in Post-Modernity, ed. D. Verducci, pp. 519-560.

Grandpierre, A. 2017a, The Fundamental Biological Activity of the Universe. Analecta Husserliana (in print).

Grandpierre, A. and Kafatos, M. Biological Autonomy. Philosophy Study 2, 631–649. 2012.

Grandpierre A, and Kafatos M. Genuine Biological Autonomy: How can the Spooky Finger of Mind Play on the Physical Keyboard of the Brain? In P. Hanna (Ed.), An Anthology of Philosophical Studies,Vol. 7, Athens Institute for Education and Research, Athens, pp. 83-98. 2013.

Grandpierre, A., Chopra, D., Doraiswamy, M., Tanzi, R. and Kafatos, M. A Multidisciplinary Approach to Mind and Consciousness. NeuroQuantology 11: 607-617. 2014. http://www.neuroquantology.com/index.php/journal/article/view/703

Hawking, S. 1988, A Brief History of Time. Bantam Books, New York, 142.

Heyneman, M. 1993, The Breathing Cathedral. Feeling our Way into a Living Cosmos. Sierra Club Books, xxiii.

Hilbert, D. 1918, 'Axiomatisches Denken', Mathematische Annalen 78, 405-415. An English translation appeared in W. Ewald (ed.), From Kant to Hilbert: A Source Book in the Foundations of Mathematics, vol. II. Oxford: Clarendon Press, 1996, pp. 1105-1115.

Hoyle, F. 1994, Home is where the wind blows : chapters from a cosmologist's life. University Science Books, 423.

Ipolyi, A. 1987, Magyar Mythologia (Hungarian Mythology, in Hungarian), 4.

Kirk, G. S. 1975, Heraclitus. The Cosmic Fragments. A Critical Study. Cambridge University Press, 40.

Kuhn, T. S. 1995, The Copernican Revolution. Planetary Astronomy in the Development of Western Thought. Harvard University Press, 6.

Lanczos, C. 1949, The Variational principles of Mechanics. The University of Toronto Press, xxiii, xxiv.

Lenormant, F. 1999, Chaldean Magic. Its Origin and Development. Samuel Weiser, York Beach, Maine, 113

Lewis, D. 1986, Philosophical Papers, Vol. II. Oxford University Press, New York, Oxford, ix.

Long, J. Bruce 1995, entry "Cosmic Law" in: The Encyclopedia of Religions. Vols. 3 and 4, ed. M. Eliade, 88.

Popa, R. 2004, Between Necessity and Probability: Searching for the Definition and Origin of Life. Springer Verlag, Berlin.

Roberts, J. T. 2008, The Law-Governed Universe. Oxford University Press, Oxford.

Spinoza 1996, Ethics, Part I, Prop. 29, Scholium. Trans: Edwin Curley. London: Penguin.

Stöltzner, M. 2003, The Principle of Least Action as the Logical Empirist's Shibboleth, Studies in History and Philosophy of Modern Physics, 34: 285-318.

http://philsci-archive.pitt.edu/archive/00000616

Tryon, E. P. 1973, Is the Universe a Vacuum Fluctuation? Nature 246: 396-97.

Westfall, R. S. 1994, The Life of Isaac Newton, Cambridge University Press.

Wigner, E. P. 1995, Philosophical Reflections and Syntheses. Collected Works of Eugene Paul Wigner, Part B. Volume VI, ed. and J. Mehra, Springer, Berlin, 322.

Whitehead, A. N. 1925, Science and the Modern World. The Free Press, New York, 5

Zee, A. 1986, Fearful Symmetry. The Search for Beauty in Modern Physics. Macmillan Publ. Co., New York, 107-109, 143.

Zhenyu Zeng 2011, Semantic criticism: The "westernization" of the concepts in ancient Chinese philosophy—A discussion of Yan Fu's theory of Qi. Journal Article Frontiers of Philosophy in China 6(1) 100 113.

A THEORY ON THE RELATION BETWEEN QUANTUM MECHANICAL REDUCTION PROCESSES AND CONSCIOUSNESS

Gerard J. F. Blommestijn

ABSTRACT

IN THE QUANTUM MECHANICAL description of nature one distinguishes two quite different processes: 1) the completely deterministic evolution of the wave function in the absence of measurements, and 2) the measurement process itself, during which the wave function reduces in a fundamentally unpredictable way to one of its components corresponding to one of the possible values of the measured observable. There are reasons to suspect that this quantum mechanical indeterminacy plays a role in the brain processes that are related to consciousness. In this article a theory is presented on the function of quantum mechanical wave function reduction in relation to consciousness.

This theory contains type identity relations between the two above mentioned types of quantum mechanical processes on the one hand and two types of consciousness related processes on the other hand, namely 1) the processing of (sub)neuronal signals in the absence of conscious observation or experience and 2) their perception and origination as conscious events.

Since the two types of quantum mechanical processes pervade all of physical reality, not only the substances and processes related to consciousness, the proposed type identities suggest a generalization of the consciousness aspects of these processes over all of nature. At the same time, it thereby unifies two groups of interpretations of quantum mechanics: the 'early' and the 'late' reduction views. The theory is in principle open to experimental investigation and suggests future neuropsychological-biophysical correlation experiments.

1   INTRODUCTION

The classical mechanical description of the physical world was totally deterministic. The future evolution of a classical system was *completely* fixed by its state at an earlier time. In quantum mechanics the situation is quite different. There one distinguishes two types of processes, the first of which is deterministic like all processes in classical physics. This first type of process is the completely deterministic evolution of the wave function in the absence of measurements. The second type of process is the measurement process itself, during which the wave function reduces in a fundamentally unpredictable way to one of its components corresponding to one of the possible values of the observed variable (observable). Only the probabilities of the different possible measurement outcomes can be predicted, not the outcome itself. This non-deterministic process is known as the reduction of the quantum mechanical wave function (or state vector). Although quantum mechanics thereby pictures a reality that is paradoxical to our intuitive, deterministic understanding of physical reality, the concepts and principles of quantum mechanics have precise definitions in terms of well understood mathematics. For the explanation of these concepts and principles the reader is referred to a general textbook such as that of Messiah (1961) or Feynman (1965).

If a measurement process acts on one of two separate parts of a system, the unpredictable outcome of the measurement may also determine the state to which the other part of the system reduces. This is the case if the parts are related by a conservation law, such as e.g. a certain value of the sum of the momenta or of the sum of the spins of both parts. The two parts may be far apart and the choice of which one of two mutually exclusive observables one measures (e.g. position or momentum) may be made at the last moment. Yet the state of the far-away part of the system reduces in agreement with the reduction outcome of the observed part. This non-locality is another peculiarity of quantum mechanics and it was put forward by Einstein, Podolsky and Rosen (1935) to show that something was wrong (incomplete) in quantum mechanics. Later experiments, especially those of Aspect e.a. (1982) and its recent 'loophole-free' version by Hensen e.a. (2015), have convincingly demonstrated this non-locality of the reduction process and have shown that it cannot be brought about by local hidden variables determining the measurement outcomes.

We must keep in mind that it is physical reality itself that is indeterministic and non-local, and that quantum mechanics only gives a description of it, a description that is as paradoxical as that reality itself. Due to an overwhelming amount of experimental evidence, quantum mechanics is nowadays generally accepted as the basic theory, not only for physics and chemistry but also for biophysics and biochemistry. There are as yet no reasons to exclude processes that are connected to life and consciousness from the competence of quantum mechanics.

Quantum mechanics may be anti-intuitive, but at the same time it frees our minds from the burden of having to explain everything, including consciousness, in terms of completely deterministic processes. For some of us this burden is felt as light or even as non-existing (see e.g. Dennett, 1993). Others think that no purely deterministic (biochemical) 'machinery', however big or complex, will be able to generate conscious understanding, creativity, will etc. If this is correct, the reduction of the quantum mechanical wave function is the only candidate principle in nature to account for such non-determinism.

The non-determinism of the reduction process, as well as its relation with observation, both point to a possible role for this reduction process in the coming about of conscious experiences. As seen above, the reduction process is also capable of generating wave function reductions of separate parts of the same original system simultaneously. This could be related to the binding together of the multitude of sense data and mental data that are consciously perceived and experienced at the same moment.

The argument of Penrose (1989, 1994), based on the theorem of Gödel, also suggests the relevance of indeterminism for consciousness. The result of that argument is that human mathematical understanding cannot be reduced to (knowable) computational mechanisms. A number of other authors have also argued in favor of a connection between the quantum mechanical wave function reduction process and consciousness. Some of them will be mentioned in section 4. A few of them suggest connections that correspond to the theory proposed in this article, but in all cases there are differences.

In the theory presented in this article quantum physics is taken just as it is, without adding any concepts or principles and without modification. With respect to the reality status of the quantum description, the viewpoint adopted here is that the wave function (state vector) not only describes an ensemble of identically prepared particles or systems, but that it also describes the single particle or system.

This theory consists of two parts. Part I is especially dedicated to our introspectively known human consciousness. The structure of this special part suggests a generalization, which is presented in part II. Both parts of the theory consist of a combination of three hypotheses that are general in scope but that are at the same time quite precise with respect to the place of consciousness in quantum mechanical reality.

2 THEORY PART I: HUMAN CONSCIOUSNESS AND QUANTUM MECHANICS

The basic elements of quantum mechanics are the two types of processes that exist in nature: 1) the completely deterministic evolution of the wave function in the absence of measurements, 2) the measurement process itself, during which the wave function reduces in a fundamentally unpredictable way.

The theory proposed in this article assigns roles to these two types of processes in relation to consciousness. Thereby it pinpoints the 'place' where consciousness could be rooted in (bio)physics. The theory consists of a combination of three hypotheses, one in the realm of the interpretation of quantum mechanics, another in the mind–body problem area, and a third in the field of neurobiology.

In the next section, an overview will be given of these three hypotheses. In the sections thereafter their meaning and implications will be elucidated.

2.1 Three hypotheses

The three hypotheses of the special part of the theory dealing with human consciousness are the following:

Ia. The quantum mechanical wave function reduction really happens as soon as enough irreversibility, degrees of freedom, energy or other difference has developed between the different superposed terms (components, paths) of the wave function. Probably this criterion only gives a chance for the reduction to happen instead of being inescapable, but in the vast majority of cases the reduction really takes place long before, or without, human observation.

Ib. Conscious experience at the end of a chain of brain processes connected to perception, and conscious choice at the beginning of a chain of brain processes connected to motor functions, happen by means of reduction of the corresponding quantum mechanical wave function in the brain. To put it more strongly: the 'entering into consciousness' (conscious perception) and the 'emergence from consciousness' (conscious choice, will) **are identical to** the corresponding wave function reductions in the relevant areas of the brain.

Ic. In human beings some highly organized and complicated mechanism of coupling between quantum level and classical level in the brain has to be responsible for the correlated (non-local) reduction of the wave function containing all sense and mental experience alternatives of a certain moment, and also for the reduction of the wave function containing all action options, including mental actions, of a certain moment.

These three radical hypotheses and their implications will be explained in the following sections, starting with hypothesis Ia.

2.2 The moment of wave function reduction

Originally, not all scientists believed that quantum mechanics was applicable to living organisms. Later with the growing amount of experimental results in quantum physics, and in quantum (bio) chemistry, confidence in the general validity of quantum mechanics increased, also for biological systems.

Let us therefore consider quantum mechanics to be applicable to all of nature, including life and consciousness. Then every process in nature belongs to one of the earlier mentioned two types of processes, namely 1) the deterministic evolution of the wave function, or 2) the indeterministic measurement or reduction process.

A key question in the debate on the interpretation of quantum mechanics is: when does the wave function reduction really happen? There have been three main groups of answers to this question: i) at the moment of conscious observation, ii) as soon as enough degrees of freedom have been triggered or a large enough energy difference or other criterion of difference between the superposed possibilities of the wave function has been reached (e.g. the possibility of exposure of an area on a photographic plate, or the possibility of creation of an electronic pulse in a particle detector), iii) never; this is the case in the many-worlds interpretations (Everett, 1957, and many others who followed). In these types of interpretation there never happens a reduction process.

The strict consequences of the first answer – the late reduction view – were put forward by Schrödinger (1935) in his famous cat example. A cat is put in a box with a bottle of poisonous gas. The bottle will be broken by some device as soon as it is triggered by a radio-active particle. The cat will then die. At a certain moment the particle wave function is a superposition of 50% particle at detector and 50% particle not at detector. If the wave function only reduces due to measurement by a human observer, then the Schrödinger cat would be in a superposition of being dead plus being alive as long as no human observer had looked. In *The Emperor's New Mind* Penrose (1989) described a somewhat modified variant of this Schrödinger cat thought-experiment to stress the unsatisfactory status of the reduction issue. He proposed an energy criterion based on gravitation that would determine the moment of wave function reduction. In his more recent books he gives a new version of this criterion (Penrose, 1994, 2005). An experimental test of this criterion is under way (see Marshall et al., 2003).

Although the criterion of Penrose is not generally accepted, it belongs to the same early reduction category of views (the above answer ii) as the generally accepted view on wave function reduction, which considers the wave function 'early reduced' in practical, irreversible situations without the necessity of human observation. In both early reduction views it is believed that the wave function indeed reduces as soon as some criterion of difference in degrees of freedom, or irreversibility, or energy is reached between the different superposed possibilities (terms, paths) of the wave function.

Having a reduction criterion does not necessarily imply that the moment of reduction is *dictated* by that criterion. Perhaps only the *chance* for the reduction to happen is determined by the particular difference between the superposed possibilities and not the precise moment of reduction. So perhaps the point in time that the reduction takes place is also a matter of probability and as indeterministic as the outcome of the reduction. It would then be appropriate to speak of the lifetime and decay of a superposition.

In any case, there is in these early reduction views almost no chance for a superposition of a dead and an alive Schrödinger cat, because there is, relative to the criterion, an ample difference in amount of energy or degrees of freedom between the part (term) of the wave function corresponding to the particle in the detector and the other superposed part(s) of the wave function. So there is according to the early reduction views an overwhelmingly large chance that the reduction really occurs, irrespective of the observation by human consciousness.

Hypothesis Ia of the preceding section falls in the category of the early reduction views.

The next section will treat consciousness, experience and choice in relation to wave function reduction (hypotheses Ib and Ic).

2.3 'Solution' of the Hard Problem

Let us follow the path of the information from a quantum mechanical experiment in the direction of the brain activities, whereby the observer becomes consciously aware of the measurement outcome. Quantum physics not only governs the experimental apparatus and its output, but also the biochemical reactions in the observer's sense organs and neurons. If we accept hypothesis Ia of the preceding sections, we have early reduction in the experimental setup, which means that we no longer have a superposition of different measurement outcomes at the moment that the outcome has been registered by the equipment.

From this moment on the situation is therefore not different from the perception of a sense stimulus coming from a normal, non-quantum mechanical object (if such an object would exist at all). Yet, the wave function of a signal that reaches the observer's sense organ (e.g. a photon reaching the eye) may be in a superposition of different possible components with respect to the observables 'measured' by those sense organs. The wave function of a photon entering the retina may, for instance, be in a superposition of being detected by a certain light–sensitive element in the retina and of not being detected by that element. To all probability there is then enough difference between the superposed parts of the wave function (according to the criterion of hypothesis Ia) to make the wave function reduce to one of the two possibilities.

In the ensuing neuronal signal activity toward and in the brain, wave function reductions will again occur at many places, according to hypothesis Ia (e.g. of the wave functions representing the presynaptic trajectories of calcium ions that may activate neurotransmitter release). These reductions are like the reductions that create recordings of quantum mechanical measurement outcomes (recordings like photographic exposures or electronic pulses), both in terms of irreversibly triggering so many degrees of freedom that the chance of undoing them is extremely low, and in terms of being relatively simple and localized.

This cascade of neuronal processes, which also includes the joining in of signals from other stimuli and from memory, may eventually give rise to the conscious perception of the measurement outcome in the case of a quantum mechanical experiment, or of the 'normal' object(s) in other cases.

Hypothesis Ic now states that in the brain there is a molecular biological structure, e.g. a network of microtubules in the neurons (see Penrose, 1994, and Hameroff, 1994), interconnected perhaps via open gap junctions between the neurons that are involved (see Penrose and Hameroff, 2011; Hameroff, 1998; and Hameroff and Penrose, 2003), that is capable of containing a wave function that represents all alternative totalities of impressions, called phenomenal perspectives (Lockwood, 1989) that are possible at a certain moment. This molecular biological structure and the wave function that it contains are probably spread out over a considerable area of the brain and the wave function is highly non-local and correlated. Its different parts represent perceptions from the different senses and experiences of various mental functions such as different kinds of thoughts and emotions.

The perception aspect of hypothesis Ib, being the central hypothesis of this special part of the theory proposed in this article, now states that the brain process that accomplishes conscious perception and experience is identical to the reduction of the above wave function representing all alternative combinations of sense data plus mental data that have the possibility of being consciously perceived and experienced at a certain moment. The becoming conscious of mental data, or in other words, experiencing them or having them, is here considered as analogous to the perception of sense data.

Of all the quantum mechanical reductions taking place in the brain, only the appropriate ones, namely the ones of the complexly organized non-local and correlated wave function in the above molecular structure, give rise to all conscious perceptions, thoughts, emotions, etc. of the human individual as a whole.

The countless reductions that have taken place in the signal processing before the final consciousness jump were reductions to states of relatively separate and localized (biochemical/physical) observables whereas the final consciousness jump is a huge simultaneous reduction to the observables of the above explained complexly organized molecular structure in the brain. We might call this structure 'reduction boundary', because it is the boundary where the quantum reductions happen that bring the perceptions, thoughts, feelings etc. from the brain to the *abstract "ego"* as Von Neumann calls it (1932, Chapter VI. The Measuring Process; 1. Formulation Of The Problem). These reductions are continuously happening in time.

The correlated non-locality of the wave function in the microtubules or in whatever structure that accommodates it, is due to the spatial extent of the structure together with some form of interaction that the signal processing in the brain needs in order to achieve the non-local correlation in the spread out wave function. The quantum mechanical entanglement and coherence of this wave function is responsible for the binding of the different parts and aspects of the phenomenal perspective to the one unity of consciousness.

The central postulate of the theory that is put forward by Stapp (1993) is equivalent to this hypothesis Ib. The supporting hypotheses

Ia and Ic, however, contain differences with the theory of Stapp. Hypotheses Ia and Ic are more in accordance with the views expressed by Penrose (1989, 1994). (See section 4.)

2.3.1 Three stages

The above conscious perception process and the preceding processes roughly fall into three stages or aspects which may be partly simultaneous. The first stage is the general neuronal processing, the understanding of which already poses very complex problems, although these have been named the easy problems compared with the really hard problem of understanding how the phenomenal experience of consciousness itself arises. At the conference Toward a Scientific Basis for Consciousness (Tucson, Arizona, 1994) and in his book *The Conscious Mind*, Chalmers (1996) made this distinction between the easy problems and the hard problem of research into consciousness.

The second stage or aspect is the preparation of a total, correlated, non–local wave function representing all sense and mental data that have the possibility of being perceived/experienced at that moment, with probability amplitudes generated by the preceding first stage. These probability amplitudes of the experience alternatives, which will possess strongly peaked spectra for many observables, determine the chances for their being the outcome of the wave function reduction in the next (third) stage. To comply with hypothesis Ic, this wave function needs to have an interface with a sufficiently reductive macroscopic environment in the brain to provide for the differences between the superposed alternative (perception) possibilities, according to the criterion of hypothesis Ia. This preparation, as well as the reduction interface or reduction boundary, requires a complex organization in the brain, perhaps corresponding to the ideas of Penrose (1994) and Hameroff (1994), namely that the microtubules of the neurons play an important role. Other biophysical/chemical mechanisms may also be responsible for this complex organization. The interface doesn't have to be localized in one neuron or in a tiny area of the brain, since the correlated wave function may be spread out over a large area by means of gap-junction connections between the neurons, everywhere encountering a sufficiently reductive classical, macroscopic environment.

The third and last stage is the correlated, non-local *reduction* of the above total wave function, whereby conscious perception takes place of the outcome of the reduction. This outcome is composed of one selected possibility for every observable of the total perception/ experience range of the reduction interface between the prepared, intricately organized perceivables wave function and its reductive environment. This reduction is not the start of more processing in some mental space, needed to generate conscious perception. No, this wave function reduction itself **is** the conscious perception. The processing done to generate mental perceivables such as thoughts and emotions also has to happen before this quantum mechanical reduction process, whereby one not only becomes aware of the sensory but also of the mental perceivables. (See also next paragraph.)

In this last stage the wave function reduction presents its outcome, so to speak, to consciousness. But what is this consciousness? Often the word consciousness is used in such a way that its meaning also includes the so-called contents of consciousness, the objects of consciousness such as thoughts, emotions, experiences etc. But in this theory, the word consciousness is strictly used in the sense of that something that *perceives* these thoughts, emotions, experiences etc. The word consciousness designates in this theory the I that perceives, not what can *be* perceived. Something that *is* perceived or experienced is not part of consciousness in this strict sense. Because mental objects such as emotions and thoughts are experienced they do not belong to consciousness in this sense. Only the I of the mind is the consciousness referred to when in this theory the word consciousness is used. This consciousness must be considered as a feature of nature, something that simply exists or is present and that serves as the receiver of the reduction outcomes, provided that the right, complex, highly organized interface between the intricately prepared wave function and a macroscopic detection system in the brain is available. This consciousness is what Von Neumann (1935) calls *abstract "ego"* (see the beginning of this chapter 2.3).

This all sounds dualistic, but the theory proposed here is to a certain extent independent from the question whether matter gives rise to consciousness (materialism) or consciousness gives rise to matter (idealism) or that both exist in their own right (dualism). The theory only discusses the interaction between consciousness and matter, given that both exist in some way (see also the paragraph Indeterminism).

This theory solves the hard problem only partly. It explains the process of becoming conscious of something by postulating its identity with wave function reduction but it doesn't explain this consciousness, this perceiving I itself. The existence of this consciousness principle or possibility may be so basic that it defies explanation.

2.3.2 Action included

In the theory presented here, it is not only perception, input into consciousness that happens by means of the quantum mechanical reduction process. Output out of consciousness, such as choices, decisions, will and initiatives for action also happen by means of reduction of the respective parts of a wave function. On this output-side, the wave function that is reduced is the wave function representing all choice and motor alternatives of the brain at that moment. The reduction of this wave function **is identical to** the conscious decision, choice, initiative for action etc. that springs from consciousness. Probably we have to consider the total wave function at the input-side, the perception side, together with the total wave function at the output-side, the action side, as one wave function that is reduced as a whole, connected to one consciousness, one *abstract "ego"* (see earlier in this chapter). Consistent with this is the introspective experience that we don't have two distinct I's, one I that perceives and one I that wants things; it is the same I that perceives and wants. The simple Schrödinger's cat *gedanken* experiment already shows how a perception pathway (detection of the particle) and activation pathway (triggering the breaking of the bottle) can both be connected to the same reduction of the wave function.

In the input (perception) to consciousness as well as in the output from it (choosing actions) we can make a distinction between the mental (thoughts, emotions, choices of mental activities) and non-mental (sensory perception, choices of physical action) parts of the total reducible wave function. (Sensory perceptions and physical movement initiatives can also be considered mental, but in the terminology of this article they are called non-mental.)

The working hypothesis here is that non-mental and mental experiences and initiatives are all related to their own specific signal pathways and/or patterns in the brain. The various parts of the overall wave function are appropriately connected with the areas in the brain that process sensory and physical action signals on the one hand and with areas that process signals underlying thoughts and emotions on the other hand, corresponding to which parts of the overall wave function will, upon reduction, give rise to non-mental and which to mental input and output respectively.

The proposed partial solution to the hard problem is: all conscious perception of sense and mental data as well as conscious choice and action initiatives (mental as well as physical), thus all input and output to and from consciousness, happen by means of reduction of the respective (non-local) parts of a specially prepared wave function in the brain which is a highly organized representation of all experience and choice/action possibilities of that moment. This wave function is continuously prepared by brain processes striving for faithful presentation of perception data and choices for action, so as to generate (often strongly peaked) distributions of superposed possibilities for internal perceivables and 'doables' (mental and non-mental). The reduction happens due to an interface between quantum level and classical level in the brain, that may be called 'reduction boundary'.

2.3.3 Indeterminism

All processing of input signals right up to the wave function reduction that constitutes their conscious experience, and all processing from the reduction towards mental and non-mental movements at the output side is governed by the laws of quantum mechanics. The working of the neurons and synapses is, however, such that they strive for robustness with respect to fluctuations due to thermal noise or effects of the quantum mechanical uncertainty relation. To a very large extent these processes can therefore be considered as deterministic. Yet, the outcome of the main, consciousness related reduction process is, like every quantum mechanical reduction outcome, not subject to any physical law. Only the probabilities of the outcomes are subject to deterministic laws. At the perception side of the consciousness related wave function reduction, this not being subject to laws may manifest itself in the form of different possible perceptions, e.g. in the case of judgements that give two or more possible ways of experiencing the same situation (e.g. the proverbial bottle that is half full or half empty). The probability of perception of the different possibilities is determined by physical (chemical) laws, but which one of the possibilities is perceived, is completely undetermined by any law. It could well be that our faculty for having control over the way of perceiving something is exactly the inside feel by consciousness of the "outside" indeterminism of the reduction processes of the perception part of the overall wave function.

At the output side of consciousness, the motor direction, also only the probabilities of the reduction outcomes are determined by laws, not the outcomes themselves. This not being bound by laws probably manifests itself in the introspective quality of free will. On first sight this association between free will and quantum mechanical indeterminism looks contradictory, because not being subject to laws is usually considered as being random and therefore quite the opposite of being determined by the decisions that we make. However the fact that the outcome is not determined by any law does not mean that the outcome cannot be determined at all. It certainly can be determined if the determining factor is not subject to any law, fundamentally free. Therefore, the indeterministic reduction from a superposition of action alternatives to one action outcome may very well simply be identical to consciously making a free choice, unbound by any law (apart from the probabilities). Then, this free wave function reduction is exactly, what is internally experienced as a choice or a decision by the consciousness for which this reduction constitutes the control over the motor signal pathways. It must be clear, however, that consciousness (or *abstract "ego"*) in this theory cannot simply be an epiphenomenon of matter without any intrinsic possibilities. First of all it has the faculties of experience and will and secondly it freely (according to the probabilities) chooses the reduction outcome of the consciousness related wave function, feeling this as choosing an action or a way of perceiving something.

The probability amplitudes that the action and perception alternatives possess before the wave function reduction, represent the conditioning, the being inclined to a certain choice. Especially if the distribution is very peaked at one alternative, the complete freedom of will may be much of an illusion.

It is important to note the difference between the association between will and quantum mechanical reduction that is often made, namely that human will would reduce wave functions of systems outside the quantum-classical interface of his or her own brain, and the association made in this theory, namely that the exertion of human will (decision, choice etc.) is identical to reduction of the wave function in this interface, this boundary.

Quantum mechanical wave function reduction should only be considered as indeterministic in the sense that its outcome is not determined by laws, and not in the sense that nothing determines its outcome. In the above discussion it is the free decision that determines its outcome, but the free decision is not determined by anything (apart from probabilities). The probabilities for the decision or choice are brought about by brain processes, but the choice itself is free and *instantaneous*. It is not a hidden variable that is present before the choice. Therefore this view is not in conflict with the refutation of local hidden variable theories by the experiments of Aspect e.a. (1982), Hensen e.a. (2015) and others.

2.3.4 Detection and activation boundary

According to hypothesis Ia a wave function has the tendency to reduce as soon as enough difference has developed between its superposed components. The wave function representing all consciousness input and output possibilities therefore needs a reduction system in the brain that generates these differences. This reduction system has to be macroscopic enough to create these differences for all consciousness observables and 'doables'. At the same time, the quantum mechanical coherence has to be strong enough to achieve a coordinated reduction, resulting in everything perceived and everything initiated at a certain moment as an outcome. To obtain this quantum coherence there should be enough isolation between the prepared wave function and the surrounding environment.

It is the complex organization of this detection and activation boundary between quantum level and classical level in the brain (and its connections to the neural pathways) that gives rise to all possibilities of our human consciousness, such as sense perceptions, motor decisions, awareness and causation of emotions and thoughts, understanding, insights, moral decisions, creativity, exaltation etc. etc. This theory doesn't claim to provide answers on the structure and the functioning of this detection and activation boundary. It only claims that there has to be such a high level organized reduction boundary and that its function is to allow the reduction process that consists of the presentation of the immense range of perceptions to consciousness and the origination of the immense range of acts from consciousness. All input processes take place before the ultimate reduction to consciousness and all output processes after the primary reduction from consciousness. Consciousness itself contains no processes.

The main problem with the idea of mind and matter dualism, namely the necessary energy transfer between the consciousness realm and the physical realm is absent here, because during the quantum mechanical reduction process no energy is transferred, yet choices and perceptions are transmitted between matter and mind.

3 THEORY PART II: GENERALIZED CONSCIOUSNESS AND QUANTUM MECHANICS

The theory presented in the first part of this article was especially dedicated to the understanding of *human* consciousness. Its central postulate is that conscious experience and decision (also including awareness and activation of thoughts and emotions) are identical to the reduction of an intricately organized wave function in the brain, representing all mental and non–mental perception and action alternatives. If for a human the input and output processes to and from consciousness are identical to the reduction of the high level wave function representing the alternatives, then the same mechanism could also apply to animals. The level of organization of the (molecular) biological interface between the quantum level in the brain and the classical level, the detection and activation boundary, will be different depending on the complexity of the nervous system of the organism. The experiential and control repertoire of such an organism will also be correspondingly different.

A straightforward generalization of the special theory of part I suggests a comparable reduction process of a wave function containing all perception and action alternatives in the brain of an animal. Particularly in the area of mental perceptions and actions such as thoughts, understanding etc. there will be a big difference with the human range of possibilities. However, this does not mean that the principle of presentation of perceptions to consciousness and of origination of initiatives for actions from consciousness by means of wave function reduction should be absent in animals. The generalization of the theory would indeed mean that animals also have an inside experiential feel and initiative, and even the same principle of I-ness or consciousness or *abstract "ego"* as humans have, but connected to a very different detection and activation boundary of wave function reductions. Simply because this reductive interface, and therefore the possible outcomes of the reductions, are so different, their internal experiential and activational world will be very different from ours (and all species from each other). In this respect it may even be more appropriate to talk about awareness instead of consciousness to emphasize the very general and not necessarily high level character of what is meant by consciousness here.

It must be noted that the inputs and outputs of consciousness/awareness are **everything** that is experienced and initiated by the organism, including self-image, reasons for choices, emotional status, beliefs, etc. So these inputs and outputs are completely determining for the kind of 'inside feel' that a certain animal or a human being experiences, while the 'generalized consciousness' experiencing the inputs and initiating the outputs, is the same consciousness or I-ness principle as that of the special part of the theory. It is that same something which experiences, only the experiences differ (see the end of paragraph 2.3.1 for an explanation of the term consciousness in this article). The adjective 'generalized' is just added to indicate that we are not only dealing with human consciousness, but with consciousness in general: awareness on all levels of the evolutionary tree of life.

This generalized part of the theory even goes to the extreme limit by postulating generalized consciousness not only over the total range of organisms but also over the inorganic systems in nature where quantum mechanical reductions take place. This is indeed a much broader range than humans and animals. However, we have no idea how the inside feel is of an animal, let alone of a plant or a piece of inorganic matter.

For living organisms there may be some form of organized reduction boundary that allows for wave function reductions that constitute the internal feel of that organism as a whole, even in the case of organisms that don't have a nervous system. The lower we come on the branches of the evolutionary tree, the more different this reduction boundary and inside feel will be from ours, due to the totally different set of (partially) organized wave function reductions, transmitting the experiences and initiatives to and from the generalized consciousness of the organism as a whole.

Stepping off the evolutionary tree, we come across e.g. quantum physical experimental setups, where according to hypothesis Ia quantum mechanical reductions also take place. Therefore, the generalized theory infers also in these cases a certain (small) amount of generalized conscious inner feel, corresponding to the relatively simple organization of the quantum-classical interfaces in these setups.

Then we have the large amount of wave function reductions occurring in inorganic nature, for instance where single particle wave packages that have become spread out in space, run into macroscopic quantities of solid or liquid matter and reduce to one of the different possible cascades of interactions. This same type of more or less isolated reductions happen in living organisms, without these reductions being part of the reduction of the total, correlated wave function at the reduction boundary that is related to the organism as a whole. Examples are the reductions of photon wave functions in the retina of the eye and the reductions of calcium ion wave functions in the presynaptic binding processes resulting in neurotransmitter release.

The existence of these and many more types of isolated reductions in the body of an organism means that not only the reduction boundary of the organism as a whole is present in it, but a lot of other instances of lower level reduction boundaries. Reductions, not related to the reduction boundary of the organism as a whole, may happen in rather independent, but yet locally organized reduction interfaces of parts of the organism, such as e.g. single cells and parts of cells, thereby constituting the inputs and outputs of a hierarchy of reduction boundaries to and from generalized consciousness.

The generalized theory of the relation between quantum mechanical reduction process and consciousness is formulated as generalizations of the hypotheses of section 2.1. These generalized hypotheses follow hereafter.

3.1 The 'Right' Consciousness-Quantum Mechanics Mix

As seen in part I, the basic elements of quantum mechanics are the two types of processes that exist in nature, namely (1) the completely deterministic evolution of the wave function in the absence of measurements, and (2) the measurement process itself, during which the wave function reduces in a fundamentally unpredictable way. The deterministic evolution type of process generates probability amplitudes for perceptions and actions but is not directly involved in conscious perception or conscious initiation of action itself. It is the wave function reduction type of process that is identical to the *entering into generalized consciousness* and the emerging from generalized consciousness, which gives rise to the experience of sense data and mental data as well as to the conscious choice of mental and physical actions. The meanings of the words experience and conscious choice are now, however, very general. Generalized consciousness still means that which perceives and chooses, and should not be confused with its contents, namely the things that *are* perceived and chosen, such as for instance thoughts and emotions, plans, dreams etc.

The generalized theory is based on generalized versions of the three hypotheses of the first part of the theory (section 2.1).

IIa   This hypothesis is exactly the same as its special version Ia.

IIb   The entering into consciousness (conscious perception) and the emergence from consciousness (conscious choice, will) **are identical to** the corresponding wave function reductions in the relevant areas of the brain in the case of human consciousness. Quantum mechanical wave function reductions, however, do not restrict themselves to brain processes related to human consciousness, but are all-pervading in nature (see hypothesis IIa). Therefore this identity is generalized to: The reduction process **is always identical to** the entering of a generalized experience into generalized consciousness or the emergence of a generalized choice or will out of generalized consciousness, in both directions via the macro-quantum reduction boundary of an organism or a quantity of inorganic matter.

IIc This entering and emerging in and out of generalized consciousness happens in all kinds of organisms and systems: from human beings to animals and plants and from physical experimental setups to pieces of inorganic matter. In human beings some highly organized and complicated mechanism of coupling between quantum level and classical level in the brain has to generate the differences of hypothesis IIa in order to facilitate the correlated (non-local) reduction of the wave function containing all sense and mental perception data of a certain moment, and also for the reduction of the part of the wave function that contains all action options (including mental actions) of a certain moment. Stepping down along the evolutionary tree of life we meet less and less highly organized and complicated organisms and correspondingly simpler macro-quantum interfaces or reduction boundaries. This will give rise to simpler levels of experience and choice repertoires for generalized consciousness. Even in physical experimental setups and objects of inorganic matter there is entering and emerging in and out of generalized consciousness, namely corresponding to the (correlated) wave function reductions at the relatively modestly organized boundaries between the quantum and the classical levels of the systems.

3.2 Reconciling the early and late reduction views

With the help of hypothesis IIc even the two opposing views on the moment of wave function reduction are reconciled: the particle detector in the Schrödinger cat setup possesses a (small) reduction boundary. The reduction of the particle wave function in the detector is identical to the entering of a generalized experience into generalized consciousness via the reduction boundary of the detection equipment (hypothesis IIb). In this theory it doesn't have to be human consciousness that reduces the wave function. Any instance of a macro-quantum reduction boundary coupled to generalized consciousness may cause the reduction of the wave function, and thereby establish a generalized experience (in this case the detection of a particle) and/or generate a generalized choice (in this case the triggering or not of the breaking of the bottle in the box containing the cat). It is some energy, reversibility or other criterion that determines or allows this generalized conscious observation and/or generalized conscious choice, so it is early reduction, because it happens without the intervention of human observation. At the same time it is (generalized) consciously perceived, so it is late reduction, namely by conscious observation.

The long sought for hidden variables of quantum mechanics, namely the theory elements determining the reduction outcomes, are in this article identified with decisions and choices of actions and ways of perceiving (by generalized consciousness). It is, however, confusing to call them hidden variables, because they are definitely not determined by any law, so they are elements of reality for which there is no possibility of prediction. They do not exist prior to the choice or decision. Therefore, according to the definition of Einstein, Podolsky and Rosen (1935), they cannot be considered as real physical quantities and do not help in making quantum mechanics *complete*. This entails that, if the theory proposed in this article is correct, physics will always be essentially *incomplete*. Yet, it is not randomness, in the sense of caused by nothing, that determines the outcomes of the reductions, but it is the freedom of choice of generalized consciousness that determines the reduction outcomes, via the instance of reduction boundary of the organism in question.

4 COMPARISON WITH WORK OF OTHER AUTHORS

In a letter to Born in 1926 Einstein wrote: "The [quantum] theory produces a good deal but hardly brings us closer to the secret of the Old One. I am at all events convinced that *He* does not play dice." (Pais, 1982). According to the theory proposed in this article *He* indeed does not play dice. At the same time quantum theory is indeterministic in the sense that the choices (within the probabilities) of generalized consciousness are not determined by any law or hidden variables. This is in accordance with the results of Aspect (1982) and later experiments, e.g. of Hensen e.a. (2015), that show that there are no hidden variables before the measurement outcome that determine the outcome. Instantaneous hidden variables at the moment of measurement however are not ruled out by these results. We could call them choices of generalized consciousness or of the *abstract "ego"* as Von Neumann (1932) defines it. So the Old One, God, nature, the *abstract "ego"*, consciousness, does indeed not play dice, but chooses above any law; but according to the probabilities.

Synchronous with the birth of quantum mechanics, Whitehead (1929) developed an understanding of reality that unified 'objective fact' and 'feeling' into 'actual entities' or 'occasions of experience'. They are the final real things of which the world is made up and resemble quantum mechanical reduction processes.

Stapp (1975) has described a model partly inspired by the one of Whitehead, which he later developed into a theory (Stapp, 1993, 2007) that has a lot in common with the theory discussed in the present article. The major differences are Stapp's adherence to late reduction, his less clear distinction between the contents (inputs and outputs) of consciousness and consciousness itself (that which experiences), the absence of the (bold) generalized identity theorem (reduction process *always* identical to input/output process to/from consciousness), and the absence of the idea of an organized reduction boundary between quantum and classical level.

The Orchestrated Objective Reduction (Orch OR) theory of Penrose (1994, 2011) and Hameroff (1994, 2003) on the other hand clearly emphasizes such an 'organized reduction boundary': In neurons quantum-superposed states come into existence in tubulin protein molecules. They remain coherent and recruit more superposed tubulins until a quantum-gravity threshold for the space–time separation of the superposed possibilities is reached. This pre–reduction coherent superposition ("quantum computing") phase is equated with preconscious processes. At that threshold (of gravitational self-energy of separation of mass distributions multiplied by coherence time) self-collapse, or objective reduction (OR) abruptly occurs. Each OR is equated with a discrete conscious event. Which outcome of OR happens, may arise as a result of presently unknown "non–computational" mathematical/physical (i.e. "Platonic realm") theory, not algorithmically deducible.

This Orch OR theory too has a lot in common with the theory of this article. Penrose and Hameroff, however, also do not clearly distinguish between an ontologically independent I of the mind or (pure) consciousness or I-ness on the one hand and the contents of this consciousness such as thoughts, feelings, experiences, choices, etc. on the other.

Some other authors that have published on quantum mechanics and consciousness:

Von Neumann (1932) has shown that quantum mechanics gives the same measurement probabilities irrespective of the position of the dividing line between observed system and observer, even if the line is drawn just in front of the *abstract* "ego".

Wigner (1967) proposed a way out of the measurement difficulty by postulating that the equations of motion of quantum mechanics are non-linear if conscious beings enter the picture.

Walker (1970) describes a theory in which consciousness is associated with quantum mechanical tunneling processes in and between the neurons, facilitated by arrays of propagator molecules. In his theory consciousness determines the outcome of the events related to the initiation of motor functions. He considers the universe inhabited by conscious (usually non-thinking) entities determining singly the outcome of each quantum mechanical event.

Lockwood (1989) adheres to the relative state formalism which is a no-reduction-at-all view. Yet it is for a conscious subject as though a particular state (component of the superposition) has been selected of the physical system (within the subject's brain) that immediately underlies the subject's stream of consciousness. Lockwood says of this state that it is designated by consciousness.

Finally, panexperientialist metaphysics proposes experience as an ingredient throughout the universe, permeating all levels of being. See for instance a discussion reported by de Quincey (1994).

5   OBSERVATIONAL IMPLICATIONS

It is rewarding to have a theory on the relation between reduction process and consciousness that meanwhile reconciles different interpretations of quantum mechanics, but can it be tested? The answer is: in principle yes. Because a wave function that has undergone reduction to one of its components differs from the original superposition, it will in principle be possible to do measurements on the specially prepared wave function at the reduction boundary in the brain to discriminate between whether reduction has or has not taken place. This must then be correlated with the conscious perception/ decision that is related to this reduction.

Direct measurement at the reduction boundary in the brain, however, to know whether or not a reduction has taken place, will be extremely difficult. Presently we do not even know what the reduction boundary consists of. In addition we have the existence of reductions along the signal pathways that are not related to the reduction boundary of the organism as a whole.

For the time being we must resort to indirect measurements, such as those of Nunn e.a. (1994), where the recording or not of the EEG of a subject's brain hemisphere was correlated with task performance parameters, to see whether measuring the EEG would enhance the reduction of the consciousness related wave function (by more easily reaching the reduction criterion) and thereby influence the corresponding conscious perception or activation. The reported experimental results were positive, but have to be reproduced, and possible alternative explanations should be sought for and excluded.

One could also try to use EPR-like (corresponding to the Einstein, Podolsky, and Rosen (1935) paradox) correlations between pairs of photons, one of which is detected in a physical fashion (e.g. by a photomultiplier) and the other by the eye of a subject, in the (probably vain) hope that part of the wave function of the biophysically detected photon would keep its coherence with the other photon until the moment of conscious perception. In this way one could devise variants of quantum–optics experiments that have been done using parametric down–converters (e.g. Rarity and Tapster, 1989, 1990).

One could also start at the other side of the problem by attempting to do quantum mechanical experiments and calculations on pieces of microtubules, such as Penrose (1994) suggests, to see whether the necessary quantum coherence could actually exist.

The first experiments in this direction have been done by Bandyopadhyay et al (2011). They report lossless electronic conductance of single microtubules at certain resonant frequencies, which seems to be essentially quantum conductance.

6 SUMMARY AND CONCLUSIONS

The world of quantum mechanics has two types of processes: the predictable wave function evolution process and the indeterministic measurement process. The measurement or wave function reduction process has three properties, which are absent in classical physics and which may relate to consciousness: a) it has the standing-above-all-laws property in the sense that it is, apart from probabilities, fundamentally impossible to determine its outcome in advance, b) it is related to observation, and c) the wave function reduction is non-local, i.e. may have instantaneous distant effects. These three quantum properties of the wave function reduction process could be related to three properties of consciousness, namely: ad a) free will, ad b) the existence of an inner world of awareness and ad c) the binding problem or unity of consciousness.

In this article the following type identity is proposed. Quantum mechanical wave function reductions are *identical to* the input processes of non-mental and mental experiences and the output processes of non-mental and mental activations to and from generalized (not only necessarily human) consciousness via instances of a quantum-macro reduction interface or reduction boundary. Instances of such reduction interfaces may be connected to consciousness in human beings, but also to the experiential I-ness in an animal or a plant or to the inside feel of a living cell or an object made of inorganic matter. Within an organism there will be a hierarchy of instances of such interfaces, all coupled to generalized consciousness.

Wave function reductions in quantum physics experimental setups are a relatively simple kind of reductions. In living organisms the (non-local) reductions that are related to the experiences and choices of the organism as a whole, take place in a complex reduction boundary between the quantum level and the classical level (in the brain of the organism if it has one). The complexity of this boundary and the corresponding range of possible experiences and initiatives depend on the place of the organism on the evolutionary tree of life.

Generalized consciousness is the observer or *abstract "ego"* in the sense of quantum mechanics, which means that the wave function at a reduction boundary reduces at the moment that an experience enters generalized consciousness or an initiative/choice for action emerges from it. Because consciousness has been generalized to be the I-ness of all beings, most of the wave function reductions happen without being related to a human perception or activation. In general, wave function reduction has a tendency to take place when there is enough difference (according to some criterion) between superposed wave function components. These differences arise in the reduction boundaries of humans, animals, plants, cells and pieces of inorganic matter.

In this theory the wave function reduction may roughly be compared with passing something through a service-hatch between a kitchen and a living-room. In the 'kitchen' a lot of processing is done: all neural and sub-neural processing, all chemical and physical processing present non-mental and mental experiential and activational possibilities at the service-hatch, the reduction boundary. The quantum mechanical reduction processes then correspond to the passing through the service-hatch to the living room (the generalized consciousness) of one of the alternatives of the total perceptive range (including mental perceptions such as thoughts and emotions) as well as the passing in the other direction of one of the alternatives of the total activational range (including the choice of mental activation of thoughts and emotions). Often the probability spectrum of the alternative possibilities for a certain perceivable or doable will be very peaked, thereby making the outcome of the wave function reduction almost predictable.

ACKNOWLEDGEMENTS

I would like to thank K. van Doorn and D. Dieks for helpful discussion and comments, F. Stewart and G. Ramakers for critical reading of the manuscript and G. Vonk for converting it from Latex to MS-Word and for editing it.

REFERENCES

Aspect, A., Grangier, P. and Roger, G. (1982), 'Experimental realization of Einstein-Podolsky-Rosen-Bohm *Gedankenexperiment:* a new violation of Bell's inequalities', *Phys. Rev. Lett.*, **48**, pp. 91-4.

Bandyopadhyay, A. (2011), 'Direct experimental evidence for quantum states in microtubules and topological invariance', *Abstracts: Toward a Science of Consciousness 2011, Stockholm, Sweden (http://www.consciousness.arizona.edu)*.

Chalmers, D. (1996), *The Conscious Mind* (Oxford University Press).

Dennett, D.C. (1993), *Consciousness explained* (Penguin books, London).

Einstein, A., Podolsky, B. and Rosen, N. (1935), 'Can quantum-mechanical description of physical reality be considered complete?', *Physical Review*, **47**, pp. 777-80.

Everett, H. (1957), "Relative State' formulation of quantum mechanics', *Quantum Theory and Measurement* (ed. J. A. Wheeler and W. H. Zurek, Princeton University Press, 1983). Originally in *Rev. Mod. Phys.*, **29**, pp. 454-62.

Feynman, R.P., Leighton, R.B. and Sands, M. (1965), *The Feynman lectures on physics,*

Vol. III (Addison-Wesley Publishing Company, Reading, Massachusetts).

Hameroff, S.R. (1994), 'Quantum coherence in microtubules: a neural basis for emergent consciousness?', *Journal of Consciousness Studies*, **1**, pp. 91-118.

Hameroff, S.R. (1998), 'Quantum computation in brain microtubules? The Penrose-Hameroff "Orch OR" model of consciousness', *Transactions of the Royal Society (London) Series A*, **356**, pp. 1869-1896.

Hameroff, S.R., and Penrose, R. (2003), 'Conscious Events as Orchestrated Space-Time Selections', *NeuroQuantology*, **1**, pp. 10-35.

Hensen, B., e.a. (2015), 'Loophole-free Bell inequality violation using electron spins separated by 1.3 kilometers', *Nature*, **526**, pp. 682-686.

Lockwood, M. (1989), Mind, *Brain and the Quantum* (Blackwell, Oxford).

Marshall, W., Simon, C., Penrose, R., and Bouwmeester, D. (2003), 'Towards quantum superpositions of a mirror', *Physical Review Letters*, **91**, pp. 13-16.

Messiah, A. (1961), *Quantum Mechanics, Vol. I and II* (North-Holland Publishing Company, Amsterdam).

von Neumann, J. (1932), *Mathematische Grundlagen der Quantenmechanik* (Springer-Verlag, Berlin). (Translation by R.T. Beyer (1955), *Mathematical Foundations of Quantum Mechanics* (Princeton University Press).)

Nunn, C.M.H., Clarke, C.J.S. and Blott, B.H. (1994), 'Collapse of a quantum field may affect brain function', *Journal of Consciousness Studies*, **1**, pp. 127-39.

Pais, A. (1982), '*Subtle is the Lord...*' *The science and the life of Albert Einstein* (Oxford University Press).

Penrose, R. (1989), *The Emperor's New Mind* (Oxford University Press).

Penrose, R. (1994), *Shadows of the Mind* (Oxford University Press).

Penrose, R. (2005), *The Road to Reality* (Vintage Books, London).

Penrose, R. and Hameroff, S.R. (2011), 'Consciousness in the Universe: Neuroscience, Quantum Space-Time Geometry and Orch OR Theory', *Journal of Cosmology*, **14**, pp. 3-39.

de Quincey, C. (1994), 'Consciousness all the way down?: An analysis of McGinn's critique of panexperientialism', *Journal of Consciousness Studies*, **1**, pp. 217-29.

Rarity, J.G. and Tapster, P.R. (1989), 'Fourth-order interference in parametric downconversion', *Journal Opt. Soc. Am. B*, **6**, pp. 1221-6.

Rarity, J.G. and Tapster, P.R. (1990), 'Experimental violation of Bell's inequality based on phase and momentum', *Phys. Rev. Lett.*, **64**, pp. 2495-8.

Schrödinger, E. (1935), 'Die gegenwärtige Situation in der Quantenmechanik',

Naturwissenschaften, **23**, pp. 807-12, 823-8, 844-9. (Translation by J.D. Trimmer (1980), In *Proc. Amer. Phil. Soc.*, **124**, pp. 323-38.)

Stapp, H.P. (1975), 'Bell's Theorem and World Process', *Nuovo Cimento*, **29B** (2), pp. 270-6.

Stapp, H.P. (1993), *Mind, Matter, and Quantum Mechanics* (Springer-Verlag, Berlin).

Stapp, H.P. (2007, 2nd Edition 2011), *Mindful Universe: Quantum Mechanics and the Participating Observer* (Springer-Verlag, Berlin).

Walker, E.H. (1970), 'The Nature of Consciousness', *Mathematical Biosciences*, **7**, pp. 131-178.

Whitehead, A.N. (1929), *Process and Reality* (Macmillan Publishing Co., New York). (Corrected edition (1979) The Free Press, New York.)

Wigner, E. (1967), *Symmetries and Reflections* (Indiana University Press, Bloomington).

DIRECT EXPERIENCE

The open door to realize limitless consciousness

Klaus Stüben

THE DILEMMA OF SCIENCE INVESTIGATING CONSCIOUSNESS

CONSCIOUSNESS IS MOST OFTEN believed (consciously or subconsciously) to be something like a power that is inside us. This is a common belief among the majority of all people independent of their background. Depending on our goals we need to unwind or develop it, find an access or approach it in ways that support us in what we like to achieve. And no matter what goals we follow, in order to reach, it is up to us to control that force, change our mind...etc.

What if consciousness is not something inside us, like a force that we have and need to wake up and use, rather something that we are a part of. What if there is nothing else but consciousness, no matter what is showing up at any moment. What if we are also only showing up in that consciousness as a product of consciousness – showing up in whatever form, whatever feeling, whatever emotion and sensation. What if consciousness can only be experienced in a direct way. As it will be shown in case examples in this article as well as experienced in different experiments the qualities of consciousness can also be experienced directly.

Consciousness is a topic of research that people have investigated for thousands of years and described in many ways. Over the years it has been mostly religions that spread out the seeds of what they realized. And often that has been used for manipulations in many ways. Nowadays it is modern science that tries to give an objective perspective of what is called consciousness. It is the presupposition of science to find an objective view to whatever is investigated so that everybody is able to comprehend the results without any need of beliefs and concepts. In practical this means to step outside yourself and take a look from that outside point of view on the investigated topic. This is the standard procedure in science and it is successfully accomplished in many experiments. This is useful in order to understand the regularities of mechanics, but doomed to failure if it comes to understanding very subtle intricacies and (indeed) useless when investigating consciousness itself. Because this means to divide consciousness into parts, one part is consciousness itself, which is observing what is happening in any investigation, and the other part is consciousness that is being looked at. So this procedure is not working and modern science is stuck with this dilemma. As the scientists are also not automatically free from the belief of a consciousness inside of themselves there are even more impediments to the experiment. It is the idea of an objective observer that is observing and not influencing the experiment that has also been realized as an untenable theory in Quantum mechanics since the early 20th century. (Wheeler 1974)

As I mentioned before, the access to consciousness could be a more direct way and the scientific approach is not so much only a scientific method as also a quite normal strategy for everyone in the world dealing with everyday life and leading to a behaviour that is quite common. It is the procedure that most – not to say all– of us use in order to deal with the everyday life and create their own world. By stepping outside of ourselves and having a look from that outside point of view to what we call our lives. This stepping outside is very often experienced as looking from somewhere behind the eyes and between the ears to what ever shows up. This imaginary point is misidentified as the source of "me" and what we are looking at from there is the world – somehow outside. As in the scientific world by stepping outside to look at the topic of research we are separating consciousness into observer and something to observe. This is called duality and it is both the source of knowledge and insights and the cause of suffering when we identify ourselves as the imaginary point behind the eyes and between the ears. So from that point of perspective – in that distance between the "me" and the world outside - we consequently feel separate and somehow unfulfilled, imperfect, incomplete or in whatever ways it shows up individually for each of us. In the attempt not to sense this, we most often try to change, control or hide that feeling in any possible way, because we believe that to be our nature. But actually, these feelings are the logical effect of the separation and not the source of our being. (Stüben 2010)

The origin of that is not only the initially mentioned belief of the consciousness inside of us, but also the fact that we have mostly not investigated what we call "me" or "I", which in a pure inquiry turns out to be a product of consciousness. If we take something for granted, which is a product of consciousness or just an idea, instead of looking deeper (more focused) we are all stuck in the same dilemma that science is.

This dilemma is not going to be solved in the same way that it shows up – by stepping outside, separating and identifying.

The way to solve it is going into the depth of what we perceive – to directly experience.

PUTTING OURSELVES INTO TRANCE 1 BY VEILING PERCEPTION

By looking of what it is that we take for granted in everyday life, most of us agree that there is a world, an environment which surrounds us and gives us the possibility to live in, move and act. Somehow we are separated from the world. We are a part of the world but separated – one by one. This is the human condition looking from that outside (objective) perspective. It is a presupposition that is based on an idea that we have learned to take for reality. If we really investigate the reality of the world that appears solid and real like Einstein, Heisenberg and others have described we have to admit that it exists of 99,999 % empty space – nothing. Depending on how deeply we look at it we could be aware of the overall appearance or focus on detail. Something that appears to be solid and materiel disappears in the ocean of empty space, of nothing. No matter what it is you are looking at, it is made of mostly nothing. But how do we create a world out of that nothing – a world surrounding us and giving us the possibility to live in, move and act? A world that without any doubt appears real! A world that we can see, feel and react to. How could something that consist of 99.999% nothing appear real? (Gribbin 2009) What we call reality is a perception that is made by our senses of what we see, hear, feel, taste. Only all aspects put together with physical sensations, emotions and thoughts seem to be real.

Imagine yourself watching a movie (a picture stream without tone) and listening to a radio at the same time. You are watching pictures, but they make no sense. Somehow it is not real. But if you switch off the radio and put on the sound of the movie, you can experience the movie and you somehow get involved. Emotions will automatically show up due to the action on the screen and although it is obvious that it is just flickering light with background noises that match perfectly to what you perceive. Suddenly the scene becomes real – light and tone fit together, the matching emotions arise and you are taking part in the movie.

This is what happens to us every single moment in our everyday life – to all of us. We put ourselves totally in trance. We live in an illusion that we believe to be true and we are both the hypnotist and our own subject.

By veiling the perception, giving meaning to it and finally call it our personal life.

If we investigate bit by bit into what we really perceive with our senses, we can step out of the illusion. We can slow down the speed, take apart the components of that trance and realize the empty space in all single parts of it and directly experience the consciousness that all this is showing up in as silence, emptiness and bliss. By directly experiencing what we perceive with our senses without giving any meaning to it, the presupposition of a world outside and the "me" inside will automatically disappear. The only requirement for this is the willingness to give up the identification with the imaginary point behind the eyes and between the ears as our selves. This point is absolutely illusionary and life itself happens not according to that perspective and can only be experienced directly.

Wei Wu Wei (1982) brought this to a point in his book "Open Secret":

"As long as a subject is centred in a phenomenal object, and thinks and speaks from there, the subject is identified with that object and is bound. As long as such condition remains, the identified subject can never be free – for freedom is liberation from that identification.

Abandonment of a phenomenal centre constitutes the only 'practice', and such abandonment is not an act volitionally performed by the identified subject, but a non-action leaving the nouminal centre in control of phenomenal activity, and free from fictitious interference by an imaginary 'self'.

Are *you* still thinking, looking and living, as from an imaginary phenomenal centre? As long as you do that you can never recognise your freedom.

Could any statement be more classical?

Could any statement be more obvious?

Could any statement be more vital?

Yet – East and west – how many observe it?

So

Could any statement be more needed?

Note: 'Wu Wei' merely implies an absence of volitional interference.

Whom do I mean by 'you'? I mean 'I'. I am always I, whoever says it, man or monkey, nouminally or phenomenally, identified or free - and there is no such entity".

Experiment for direct experience, no 1:
Localise the point or space that you are referring to when you say "Me" or "I" and apply only to what you directly experience without falling back to what you know, what you have heard, your memory or any presupposition. What is it that you are aware of there? What do you really perceive?

DIRECT EXPERIENCE

In the famous Yoga-sutras of Patanjali (1976) he describes the discipline of yoga as a state of consciousness when all mental activities come to rest. "Yoga is the restraint of mental modifications". In that the observer (seer) rests in the essential identity. All other internal conditions are determined by identification with mental modifications. Patanjali mentioned the following 5 mental modifications, no matter of being experienced as suffering or not: valid knowledge, error, imagination, sleepiness and memory. The actuality of these words is still given although they have been written 2500 years ago. These notes of mental modifications describe accurately the origin of what I call trance or illusion, which is a state of consciousness that is showing up and disappearing again. (Stüben 2010)

Because of our direct experience – with its imaginary point of perspective that we are identified with and call "I" or "Me" - is followed by a cascade of emotions, feelings, and thoughts with some relevant meaning to our perception. We are used to react to that in our individual way and try to control, avoid, cover up, change or hide the reaction as we believe that is who or how we are and take the misidentification for real. All our attempts portray our personal story that we call our life and call by our name. The whole lot of reaction patterns we call our personality. No matter whether we express or suppress our emotions, both ways are based on the presumption of a separate "I" in the centre of the perception. In both ways the direct experience is veiled with the effect of being separated and feeling somehow unfulfilled, imperfect or incomplete due to that distance.

In our everyday life, this mechanism of separation and identification by stepping out is subconsciously running automatically and undermine our whole life. And by investigating this very mechanism of separation we can also use this in order to understand how we create our world and what it is, that we need to do in order to imagine it being real. Investigating means to directly experience in the meaning of that word – direct, new, fresh, free without any presupposition. This comprises all aspects of life – physical, mental and emotional and is free from wishes, imaginations, knowledge, belief, preferences, aversions or limits. The tendency to react to what we perceive is also welcomed to be experienced without giving any meaning to that. There is no separation in direct experience but instead the observer and the observed (object) melt into one in the moment of observing as an energy of oneness that fully integrates physical, mental and emotional aspects. Those aspects will only be sensed as separate and partitioned in the illusionary separation of observer and the observed (object) if we choose to identify with the illusionary point. Direct experience is like gliding into a lake – melting with that which we experience. In direct experience we are inside the centre of perception. Very often while working with people, I experience that those people separate and identify themselves again very subtle. They report of pictures that appear in front of them. Giving up again this perspective is direct experience.

We can only experience directly but not really describe it with words, because we all tend to give meaning to what we hear and every word is a separation from oneness and do not report the fullness of experience.

Some people try to describe their experience as a coming home or a supernova of expanding joy, that is too big for the body. Others perceive it as roaring silence, bliss, limitless endless joy, peace and unconditional love in the heart that does not need anything else.

Direct experience is absolutely easy and naturally without any action or effort from us out of our self-induced trance. By opening into direct experience we can stop the cascades of emotions and meanings in order to reach deeper layers of experience. Like falling into something that is not really describable with words but known as "Self", unlimited endless happiness, silence and wideness might show up. There is no end in that sensation of falling and in that intensity we can experience everything that is showing up, like for example our problems as a source of happiness. If for example we sense that we are not good enough and examine that by putting aside the meaning (which is a mental aspect), we will quite likely find for example joy or power. We can now trust our experience just the way it shows up. We can also welcome our tendencies of refusal or rejection as a possibility of deeper inquiry. We can direct experience and inquire into all aspects of our life, of how we individually and specifically veil our original experiences and call this veiling our personal problem.

WE ARE CONSCIOUSNESS ITSELF

According to Ramesh Balsekar (2009) we are consciousness itself. "We have never been anything else. Maybe we understand truth better, if we tell us that there was never a "We" and that consciousness is all that was or will be. We consciously or subconsciously consider us to be sentient and in that sense special: We are the subjects and everything else in the world of manifestation consists of objects. In fact we are also just phenomenon's, parts of the universe of manifestation We consider ourselves to be special and separate because the universe notifies itself throughout our perceptual ability. Therefore, we could hardly free ourselves from that deep feeling of "Me" being different from everything else showing up in the word of manifestation. The illusion of that is, that we do not consider ourselves collectively as one perceptual ability, that realize all manifestation showing up in consciousness, but consider ourselves as separate individual beings. This is the reason for our suffering and bondage. As soon as we realize to be consciousness itself instead of separate identities the illusion of separation, which is the cause of all suffering and bondage, will disappear".

There seems to be a longing in everybody to recycle and identify again. This is also known for ages and written down in the Bhagavad-gita (1955), the holy book of Hinduism. "One develops attachment to sense objects by thinking about sense objects. Desire for sense objects comes from attachment to sense objects, and anger comes from unfulfilled desires.

It is only those who give up all desires and wander free from hankering, false ego and possessiveness who attain peace".

People tend to again and again successfully put themselves in trance and look for solutions of their problems by living that trance. Often they tend to manipulate the trance to get a better one and there do exist many offers from the outside world with good intentions. But as long as there is the false identification of a "Me" that observes, that is separated from the outside world, those offers cannot solve any problem.

Here is a brief look of how that trance of an individual identity shows up in everyday life. (Tolle 2004) As a baby we all live directly and we only learn later in life (around 18 month) to equate our direct experience with our name that was given to us. So the baby who experienced life fully begins to call himself for example John. Later on in life the baby learns that John is a boy. He learns that John is a good boy and so on… whatever attributes are added to that is veiling the direct experience more and more and with the growing belief that these attributes represent what we are, we start to be more and more identified with that implementation which is just an idea. The identification is very strong because we are convinced that all this is what we are. It is impossible to just get rid of that conviction because the one who wants to get rid of that is the one who believes to be real. The only way to get out of that is to really question that conviction and experience directly that there is no "Me" and that "Me" is an illusion, but still there is something we could call presence – silence – happiness. This is what is called the self.

Carlos Castaneda (1995) reports that it is the "inner dialogue", which is the summery of what we tell us internally, that is the source of the identification. By stopping that inner dialogue we will experience, that our identification as well as the world around us will stop to open to something new – fresh – endless. This is the same advice in all spiritual traditions. With all the experiences I gained working with people I realized that people in their willingness to following the advice to stop, are facing tremendous problems as long as they are identified as the one who stops that inner dialogue. Since in the moment of stopping there is automatically no identification, no "Me" that is involved and later they cannot remember because of the absence of identification. When that dialogue has stopped there is the one who did that stopping of the inner dialogue and is even stronger identified as someone who masters to control. It is much easier to find the source of the inner dialogue by directly realizing which is there.

Again in the Yoga-Sutras of Patanjali (1976) he describes the observer or the seer, who is that what we call "I" if there is an identification as follows:" The seer is nothing else but the energy of seeing. Although he is totally pure he sees throughout the experience. The seen exists only for the seer. For those having finished the seen is dissolving but it still does exist for the others because of the common experience".

With his own words Patanjali (1976) is describing direct experience clearly and concisely. It is really just energy that we perceive. What we make out of it, the meaning we add the seen is only important for us and it will automatically disappear, when we stop to add that meaning to our perception. Then the "seer" and "the seen" will be "One" in the seeing itself.

The traditional Zen narrative contains the well-known story of the contest for the succession of Huineng. (2006)

Huineng won this contest, but had to flee the monastery to avoid the rage of the supporters of Shenxui.

The contest was centred on two verses of Shenxiu und Huineng:

Shenxiu:
The body is the bodhi tree;
The mind is like a bright mirror's stand.
Be always diligent in rubbing it—
Do not let it attract any dust.

Huineng:
Bodhi is fundamentally without any tree;
The bright mirror is also not a stand.
Fundamentally there is not a single thing—
Where could any dust be attracted?

This text from that Zen tradition represents so concisely the problem that is automatically arising when there is misidentification, when we follow the concept of a separate individuality. This concept is leading to the origin of the idea that each individual is the source of consciousness or the mind is producing consciousness. Consciousness is everywhere and everything arises out of consciousness. That includes the idea of a separate identity. It is consciousness that causes the mind to show up.

This tendency for conceptualizing is quite strong in our western world and seems to be a spontaneous automatism in all of us. But it is a mechanism that we all learned and if we break up this mechanism in peaces, bit by bit and bring our awareness to each of them one by one, we will realize that we do need to bring up a lot of effort to put it together and make it seem real. Therefore the direct experience is a doorway to consciousness – to open up into consciousness directly. Everybody can start here and now, take what they are aware of now. The first step is to tell the truth about, what you are aware of now and what you perceive physically, what you perceive emotionally and what you perceive mentally and be clear about the assignment.

There was once a young mother consulting me, she was frustrated and desperate about her son, who was going to be thrown out of school. She was frustrated because she thought she failed as a mother and could not guide her son. She felt guilty. By directly experiencing that guilt and not avoiding it by going into that story of a failed mother, she could realize the energy that was the driving force behind it. This force suddenly was showing up as a limitless lightness and oneness with what had just before been seen as a problem.

Experiment of direct experience, no 2:

With eyes closed ask yourself the following questions without using a memory, an imagination or something you heard, which only refers to your momentary perception:

How big are you? What form do you have? Could you have almost any form? Do you feel boundaries? Is there any boundary where you end and the word begins or is there nothing separating you from the world?

You may probably perceive a variety of sounds from far away and near by, but do you perceive a sound where you are? Or what is there? Could there be silence and sounds arise and disappear out of that?

Could it be that you are the centre of accumulation of thoughts and feelings, the mind?

What and where is the mind? Do thoughts and feelings arise out of the mind or out of nothing? Is there a centre of the mind, locked up, separated from the world or a part of it, one with it?

If sensations of heat, discomfort, breathing or pleasure arise, do they make "A something" in your centre, somehow separated or do these sensations arise and disappear in the consciousness as well as thoughts, feelings and sounds (after Douglas Harding, 1988)?

A DEEPER LOOK AT WHAT WE ARE IDENTIFIED WITH

Looking into detail we realize that we subconsciously and habitually veil our direct experiences and generate distance. Realizing that as an active effort that we habitually do, we can free us from doing it – if this is what we really want. To be clear at that point it is a necessary action to divide consciousness into two aspects to learn, make experiences and understand. Realization is only possible in that polarity. It is a state of consciousness that is a requirement for learning and growing. And it is common to identify with one part of the polarity and avoid the opposite. But in identifying with one part we automatically generate distance. It is our choice but somehow we are addicted to choose one part and avoid the opposite. Both are showing up in consciousness.

If we bring our experience into words we always relate to three aspects. These are the physical, the mental and the emotional aspect of consciousness. All three aspects are showing up in a package but the mind is splitting it up into these three parts. (Jaxon-Bear 2001)

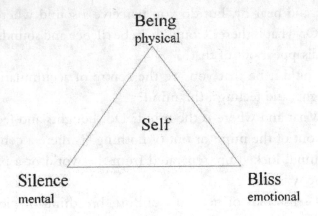

Fig. 1 Klaus Stüben

This is what we are referring to when bringing our experience into words.

If there is no identification consciousness shows up as presence (being) – silence and bliss (or whatever words there are apposite for each of us). The direct experience is perceived as unlimited, free.

In the direct experience there is no identification as somebody as an observer. So for example in the moment of being in love there is love everywhere – unlimited - no boundaries – oneness. This is our own nature and is called "Self" (Maharshi 2006) and has nothing to do with the partner that you are in love with. There is just love and as soon as there is identification there is a "me" that is going to take possession of that love by calling it "my love". This is the moment when problems begin to show up. Suddenly there is identification and separation. There is a "Me" and a "you" with all interactions in between. Instead of the physical being there is "my body", there are "my thoughts" where there was silence and instead of bliss there are "my emotions".

As soon as there is identification with a somebody as the centre of perception that called himself "I", the "self" is veiled and the three aspects of consciousness are turning into:

My body instead of pure being
My thoughts instead of a silence
My emotions instead of bliss.

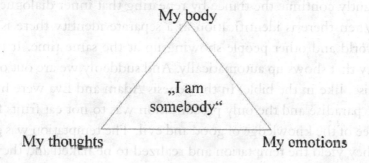

Fig. 2 Klaus Stüben

By identifying duality and polarity is showing up with all consequences in life and with all interactions between separate individuals. It all starts with one thought "I am", that causes the illusion of boundaries and keeps itself alive through the running inner dialogue. (Maharshi 2006) It is this inner dialogue that is always telling us who we are and what we need to do in order to organize our life. This inner dialogue is located at the centre of our speech in our left half of the brain and it is his job to define all details of our life. It is permanently repeating details about our life and our memory and is the home of the "ego". Without that centre we would forget who we are and what our story is about. By connecting the cells of that centre it is evolving by itself during each life. (Taylor 2009)

Stopping that inner dialogue it is not something that is an intensive desire for the ego because it will certainly mean to expose the ego as an illusion. For the ego this is a death experience and this is something to be avoided. As in all spiritual practises there is an advice to stop the inner dialogue to experience the fullness of consciousness. But because of the reasons shown above it is impossible for the ego to stop. The ego is the only obstacle to that. But – if we open ourselves to direct experience by inquiring what we really perceive, we will automatically experience consciousness directly which shows up as silence, emptiness, wideness, endlessness and bliss.

We do not need to give up anything in order to experience consciousness directly – even in duality of identification it is consciousness we are aware of – it is only to stop inducing and constantly continue the trance by repeating that inner dialogue.

When there is identification as a separate identity there is also the world and other people showing up at the same time. It is the duality that shows up automatically. And suddenly we are out of the paradise, like in the bible. In the genesis Adam and Eva were happy in the paradise and the only precondition was to not eat fruits from the tree of the knowledge of good and evil. The temptation was great and they yield the temptation and realized to be naked and they felt ashamed and guilty. They were out of the paradise and immediately were longing to come back.

The fruits of the tree of knowledge of good and evil are a synonym for duality and by eating them we are thrown out of the paradise. It is the duality in consciousness showing up and in choosing just a part of that by selecting one side of that upcoming polarity it ends up in separation – being naked, feeling ashamed. This is what the entirety of the human conditions is about that is accompanied by the desire to come home again.

When identification shows up there will be relations or reactions between "me" and "you" or "others" showing up at the same time. There are some patterns as tendencies to react that follow consequently out of the identification. In order to survive, a separated "I" will perceive the world or the others automatically as friend or as enemy/prey and react spontaneously, which means that this is driven by impulse without any conscious intention. It does not mean that there will not be reasons found for the reaction and justifications but these come up only in retrospect. What we believe to be our very own individual reaction is once we inquire into an automatic drive that is triggered by the identification.

The result is that an individual person can either flight or fight the enemy in order to survive. If the other person is concerned to be friendly or we need him personally, our reaction patter will of course drawn towards the other person. This are reactive pattern "moving against" in a case of fight, "moving away" in a case of flight and "moving towards" in a case of a friendly approach if somebody want something from the other person. It is an energetic reaction just like electrons flowing from negative to positive poles. (Stüben 2010)

But experienced from an individual point of view these reactions show up as follows:

The reactive pattern of **moving against** is causing the impulse of getting angry, in order to keep control and to be right. Out of that there is the impulse to move against. It is a very physical impulse of energy. Of course there will also be a mental aspect in form of an inner dialogue such as:" Where is the enemy! Here is the weakest part to attack?"

The reactive pattern of **moving away is** showing up as the impulse of becoming afraid, in order to think about, what to do and to find security. The impulse to move away leads to dissociation of the physical and emotional body that is energetically moving inside.

The inner dialogue may be: "Help! How can I be secure? How can I get out of here?"

By **moving towards** somebody, the reaction of becoming emotional shows up. It is like looking for love that is provided only from other people. That leads to the energetic movement towards the other person, accompanied with the attempt of attraction and forming a good image. It is energetically driven like a suction, a filling up by the energy of the other person with the following inner dialogue: "How could I be attractive to others?"

These three reactive tendencies are the result of an identification as a separate ego and not so much the result of a free and spontaneous decision. These reactions will only be interpreted later and justified by the ego as a following up of conscious decisions, and justified also by the individual history.

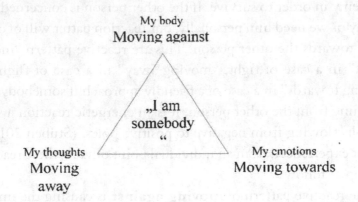

Fig. 3 Klaus Stüben

These reactive tendencies or patterns are of fundamental nature and lead to addiction, phobia and neediness to different extent. They are showing the basic energetic quality of every separated individual. They are shaping the character at the deep root of identification, the illusion of "Me" and are described in the Enneagram. (Jaxon Bear 2001; Almaas 2004; Stüben 2010)

If we just ponder for a while what kind of interaction shows up between all these different individuals it is obvious that everybody is hallucinating the own reactive tendency as the correct one. Even if every single individual has the best intention for the own life and for the relation to others there will be trance versus trance showing up that is determining all interactions. All human life on earth is determined by those reactive tendencies and we all use one of these patterns to deal with our everyday life in the everyday world.

> Experiment for direct experience no 3:
> With your eyes closed ask yourself the following questions without using a memory, an imagination or something you heard off and only refer to your momentary perception:
> "Who am I if there is no past and no future?"

With this experiment we can directly experience what we are. Putting it in words again is much more complicated.

Another very efficient experiment to realize the consciousness of what we are is the realization of what we are not. Every single word has different meaning to each of us and words are mostly not suitable to transmit experiences. But in order to describe who you are not, words are perfect. With other words: "Everything you can describe is something that you are not!"

> Experiment for direct experience, no 4:
> Write down in a list what you believe to be. Or write down one day long where and when in your everyday life you name an experience as "Me" or "Mine".

So what is it exactly that you give the name "Me" or what you describe as "I". Write down everything that is related to you. Like a detective you can draw a picture of whom you want to inquire.

Everything counts, every memory, every desire, all sensations, all thoughts and every emotion – everything that you name with "I" or "Me".

Then take that list and look at the details that you in its whole call by your name. Be honest, does that list in any way describe who or what you are? Could you find any continuity in that list that you could mark as "I"? Or is that what you call "I" much more a vague idea of something you have not inquired so far?

The practice of direct experience – waypoints and gates expected to show up

Many people misunderstand direct experience to be a technique or a method to lift the veil of illusion and experience consciousness directly. But this idea of direct experience being a technique is a trap. The ego gladly welcomes it in order to include this into the idea of a separate identity and will use the technique to get out of the trance of "me and my life". Therefore, the idea of direct experience will strengthen and solidify the illusion itself.

Furthermore, in supporting and guiding people in their direct experience I realized that most of the them have tremendous troubles or difficulties to distinguish between physical, mental and emotional aspects of their momentary experience, as well as they tend to muddle up theses aspects. This is an aspect of the trance induction, which is used in order to create the illusion of separation and it is important to clearly distinguish between those three aspects. This is of even of more importance, as we want to break up that trance in pieces to inquire each of it one by one. If people answer the question of what they feel with "I think I feel..." this answer is a very mental one. It is also mental if the answer will be: "I feel that I am wrong".

As experiences show up in a package of physical, mental and emotional aspects, we can break up that package at every step of the experience. This step is always what we are aware of in that very moment – the focus can be on the physical, mental or emotional aspect. There will certainly be one aspect of higher priority and of more relevance. The relevant aspect might and probably will change during the inquiry so it is important to stay true and focused on what shows up in the moment. Thanks to my own experiences I realized that the emotional aspect is of most importance in the whole inquiry. In our brain, emotions are processed in deeper layers than thoughts or physical sensations. In contrast to that, the thoughts are processed in more outer layers, the cerebral cortex. Therefore, the emotions are out of cortex control, which is the centre of speech and the area of the inner dialog, that defines all details of our life. This is the home of the ego that will support all ideas of its existence and will filter all deeper information. (Taylor 2009) By focussing on deeper levels we touch something that is both more essential and existential.

There is a typical pattern or layering of emotions (Jaxon Bear 2001; Almaas 2004; Stüben 2010) that I saw in many investigations. This is like an emotional ladder that has to be climbed up in order to put ourselves deeper into trance. In the inquiry and direct experience of what is perceived, people climb down the ladder again in order to realize consciousness itself.

The starting point of the inquiry is our "momentary situation". It is the situation that is experienced in that moment of inquiry of all problems, fears and desires. It is the story of "Me" and "my life" as everybody is experiencing in everyday life. It is this story that we are aware of and that keeps us busy in our mind so that we somehow avoid to directly experience what we perceive. As a wonderful example there is the story of an elderly women. She is complaining that her daughter does not visit her nor does she go shopping together with her. She is complaining about her and obviously gets more and more angry. She is looking for support from her friends in that way that they agree with her that she is treated so badly by her daughter. She is looking for compassion or advice by telling that story and by avoiding that underlying emotion of anger or sadness.

If we take a closer look at that example vicariously to all other stories, we realize that this woman is clinging to the story with the intention not to feel the underlying emotion. There could be an alternative to change the perspective.

If we put attention to the sensations and emotions underneath most stories of everyday life, we will most likely find anger or sadness. This is at least what I realized myself and by supporting other people. Surely this is true for the example above. It is not crucial for the experience that it is perceived in absolutely the same way. It is just important to be aware of what happens and to be honest about that. There is no right or wrong. The direct experience is the only authority.

Maybe the experience of rage or sadness is showing up if you put aside the story or the content of that story for a moment. The common reaction to that is either to express that rage and get angry with that person or situation or to suppress that rage by eventually denying the importance of that story. Both reactions are justified by the story itself. The example above could cause an inner dialogue such as: "Why isn't she calling?" or "I don't like shopping anyway!" Both reactions are a form of avoidance. By worrying about those reasons or by rationalizing them mentally, the direct contact with, or the direct experience of that rage is avoided. What and how that rage is thought about and whether it is liked or not is all the same. It is a fact, that everybody can be aware of pure emotion without any story.

If the rage is neither expressed nor suppressed and just experienced directly – without the story and without justification or judging – there will be awareness of what is underneath that rage or what shows up on a deeper level. Sometimes it is easier to ask what remains when the rage disappears. There might be the direct experience that this rage is made of energy. This energy changes and is nourished by something deeper. Calling it deeper refers to an expression that most people use to describe as a subjective experience.

Mostly there is a deep sadness underneath that rage. There is a tendency to now weave that sadness into a new story. People might say: "I am said. It is because my parents have treated me badly!" or "I know that I am wrong." These thoughts bring the awareness up to the surface again or project sadness into the outside world. In that mental activity, sadness is going to be avoided – avoided in order to be experienced directly.

When sadness is welcomed in the sense of experiencing it directly and without a story, judging or justifying, it will disappear. Fear will show up next.

Fig. 4 Klaus Stüben

If fear is showing up, most people will hear the alarm bell ringing. Fear is something that definitely has to be avoided. On the other hand, fear is a very useful survival mechanism, and often there is fear of fear and a mental preoccupation with the thought of fear that is blocking everything. There is a lot of medication and therapy to get rid of fear. Fear is like an electric fence that prevents us from experiencing further. It is to be avoided and it is most often not experienced directly.

If there is a willingness to look deeper, without any story or imaginations, what may happen next is, that you might be surprised to realize that fear will also disappear and/or change.

The next emotion to show up is despair. Together with despair also the impulse to judge or justify shows up and the tendency to weave all into a story again. Despair is not something that is desired but the good news about despair is, that despair is at the root of identification. If there still is the willingness for direct experience, if there is no expression or suppression, despair will disappear immediately. Then there is something appearing that often is described as a black hole and as an experience of falling through that black hole. This is where identification is ultimately dropping off. It is a fall into a wideopen space, silence and blissfulness, a state that is described as Nirwana or Samadhi. It is the direct experience of being one with everything. (Jaxon -Bear 2001; Maharshi 2006; Taylor 2009)

The observer and the observed (object) are melting into one in the process of perception. This is consciousness itself – directly experienced. (Stüben 2010)

Everybody will have their own descriptions to explain that experience. Although words are not really able to express that experience they are quite similar in many reports. It is the sensation of coming home – to be closer to yourself than to your heart.

It is the diamond in your pocket as Gangaji (Gangaji 2006) describes in her book. People report that it is like a joke that they directly experienced. What they were looking for their whole life is directly underneath of what they have mostly avoided and what they have tried to escape from. The diamond they were looking for is directly hidden underneath anger, rage and sadness. They have just looked in the wrong direction all the time.

Every desperate search for happiness and all efforts are doomed to fail while looking outside of ourself. If people are looking for answers outside to answer inner questions or look to authorities for the answer to the achievements, that they really want – this is doomed to fail. This is not a concept or an attempt in order to convince people, it is more a possibility for everybody to experience directly.

In the realization of consciousness there is the ability to directly experience what is happening, if the thought "I am (name)" arises. There will be surely the experience that the unlimited, empty space is contracting and boundaries do arise. Suddenly there is a body again and a world around it with a separation between inside and outside with all the typical sensations and emotions that are accompanied with that identification triggered by the thought "I am".

There is the possibility to also experience that the thought "I am (name)" is sinking back again in the unlimited empty space (with silence and bliss) when there is no connection with this thought. At this step there is the point of decision between freedom and identity. This is the only decision point and by looking deeper there is the realization, that there is nobody there to choose. The one who has the choice is showing up not until the thought "I am" arises out of nothing.

Through identifying by choosing identity, the empty wide space will contract, the silence will be veiled and boundaries will show up again. Immediately there will be despair about that space disappearing and the sensation of having lost something very precious. The fullness of consciousness and the feeling of coming home will never be experienced again by following that direction. Because deep inside there is the knowledge of having chosen that identity and the realization of having lost something precious, there will be fear arising – fear of finally having lost it.

If we keep on following that direction we will be experiencing sadness. So there might be resignation showing up in the certainty of having lost that treasure, followed by rage that will arise. The rage is focussed against ourselves and all the others and against the circumstance. It is a last rebellion.

Since rage does not feel good and is causing problems between ourselves and others the rage is projected into the outside world and a story is woven around that. A story that causes a feeling of not being responsible for all that – this is called everyday life.

All these emotions are reactions and interactions of the identification as an "I" with the world around us that arises from just that identification. (Remember: It is the identification that creates the duality). It is the attention that goes outwards. Having arrived in the everyday life, unhappiness is experienced including the desire to get back that which is missing so much and hoping to get it from somewhere outside of us. Together with this desire that will show up, there is also the hope that fulfilling our desires will bring back that happiness again.

It is like a joke realizing, that whatever we desire so intensely is only desirable, because the light of consciousness is shining upon it.

But a fulfilled desire makes us happy only in the moment between the fulfilment and the arising of the next desire.

Because our very own nature is happiness, we will automatically directly experience happiness when all desires disappear.

As the Buddha said: "There is no way to happiness - happiness is the way!"

Gateway of thoughts and mind – the mental body

In the Western World we give an exceptional importance to thoughts in terms of their meaning in life. Thoughts are an incredible tool for knowledge and realization. All sciences are based on thoughts. Thoughts are also often responsible for suffering once they are overestimated. On the other hand thoughts even lead to a growing number of mental illnesses once they get out of control. It was Descartes (1641) who connecting thoughts directly to existence. His philosophical proposition „I think, therefore I am "(cogito ergo sum) implies, that thinking about one's existence proves - in and by itself - that an "I" does exist to produce thoughts. Indeed thoughts determine the experience in many ways and they are often shaping the inner experience of us human beings. This is true in the sense of the proposition of Descartes, but it is based on the illusory identification of a separate identity. The thoughts themselves are putting people into trance by constantly repeating them over and over again.

Although they are putting people into trance it is possible to also use them as a means of escaping that trance and experience consciousness directly.

Ramana Maharshi (2006) answered the question of whether thoughts never showed up in his head, with that he indeed realized thoughts showing up in his mind but he did not identify with these thoughts. The thoughts are not really the problem. Thoughts come and go. They become a problem only if they are labelled as "my thought" and when they are repeated again and again. In that way they become automatic.

There are many concepts of how to deal with thoughts. All concepts can be useful but regarding the desired condition they are of different value. If there is suffering coming out of bad thoughts is surely better by thinking positive thoughts. It is possible to change the mind in order to accomplish the desired condition. But often it is found to be true that people only change their mind and think better thoughts because there is the subconscious conviction of the present condition to be true. The use of any kind of mental correction is based on the idea of the reality of a separate identity and the possibility to change that. Yes it is possible to change it, but it will not solve the problem at the root of identification, if there is the desire to directly experience consciousness in form of the self.

There is much healing in changing the mind and reversing thoughts into the opposite. Byron Katie (2002) was able to work herself out of a very strong depression and also helped a lot of people to realize how their mind creates their suffering in life. It is a very effective therapy and it starts at the point in one's life, where the problem arises. This work is keeping people in the mind by changing thoughts. This mind activity is causing duality and duality is the sign of identification.

Often there is a lot of pride rising up if the mental correction was successful, because the one who was successful will be identified even more — now as a successful separate identity who is even able to even control thoughts. It is a belief that thoughts are independent and separate of the one who is thinking these thoughts, which causes separation between the thought and the thinker. This is also identification.

There is a possibility to use thoughts in a different way. The possibility is to step directly out of the identification and use thoughts for direct experience of consciousness. If the awareness is not focussed on what the thought is dealing with, but instead on the source of the thought itself, there will be the experience of an empty space with the thought directly arising out of that empty space. By inquiring the source of thoughts people seem to perceive both that empty space and that there is nothing and nobody. Nobody is there to think that thought. Instead there is presence itself although it is neither personal nor individual.

By focussing directly onto that empty wide space, the awareness arises that there is no separation at all. Subjectively, it is the direct experience that nothing is separated from your self. Everything arises out of that empty space of nothing. This is the experience of limitless freedom without judgement, precondition or boundaries.

Experiment for direct experience no 5: (Stüben 2010)

Take any thought you like (maybe a short one) and focus directly on that thought. Ask yourself from where it arises. Do not focus on the content of that thought but rather on the story within. Imagine that thought to be a postal package and that you are the postman. Your only business is to deliver that package. The content is a secret. It is your only responsibility that the package is being delivered to the address.

Examine from where this thought arises? And whereto does the thought disappear?

Follow that thought back to its source! Where or what is that source? What do you find there? What is it that you can tell about that source?

Whereto does that thought go? Follow that thought to that point, where it is disappearing! What do you find there? What are you aware of at that point?

Focus totally on that point, where the thought arises and where the thought disappears. What are you aware of in that moment? Experience it directly!

The experiment above may lead to the direct experience of thoughts arising magically out of nothing and also disappearing into nothing. That nothingness is commonly misidentified as "I" when there is attachment on the thoughts.

If the awareness is focussed on that arising and disappearing of thoughts there is a gap showing up with no attachments to thoughts and no identification at all – just silence, peace and happiness.

It is possible to also inquire the first thought in the ladder of identification, which is the "I" thought and directly experience the source of it.

Experiment for direct experience, no 6: (Stüben 2010)

Take the thought "I am (add your name)" and focus on that thought. Ask yourself from where it arises. Follow that thought back to its source!

Where or what is that source? What do you find there? What is it that you can tell about that source?

Be directly aware of the source of that thought – What are you experiencing? What are you aware of there/here?

It is quite common to assume that, what is called "I" is located in the centre of all thoughts and is producing all the thoughts. Direct experience however is not confirming this assumption in any way. In contrary of this hypothesis many reports proof that it is the thought "I am (and your name)", that creates a separate identity that does not show up in the direct experience. Instead of the effort that a separate identified individual needs in order to control all problems in the life, the direct experience will lead to the end of that effort and to the beginning of experiencing consciousness itself in form of the self.

Gateway of sensations – the physical body

In the world of manifestation it is the physical body that people often are referring to when it comes to identification. The identification with "Me" as a physical body is very strong. In that sense it is not surprising that the recognition of the physical body as "Me" will also show up in seeing an old picture of ourself.

Certainly, there is a physical body that is subjected to transience and is going to die sooner or later. On the contrary there are more aspects of consciousness that we refer to if we identify as individuals.

The physical body itself allows us to realize and to directly experience consciousness in form of physical sensations. By judging and making it conditional the direct experience is veiled. Concerning the body`s ability to experience sensations Poonja (1993) reported of the question about sex, that he has been asked. His answer was: " to make children". To the question why it feels so good to have sex he answered: "It feels so good, so that you do it". In that answer he was referring to orgasm as a sensation of overflowing energy and this sensation is feeling good (I presuppose that this is true for almost everybody regardless of what else they believe). At the very moment of orgasm the tension of all control, desires and lust is automatically released and the identification is dropping off so that consciousness itself shows up. It is the moment of experiencing emptiness, silence and bliss – the duality between the observer and the observed (object) disappears and oneness shows up. Ramana Maharshi (2006) called this experience "self-experience".

It seems like an evolutionary trick to connect the longing for experiencing the self to the drive of reproduction in order to assure the survival of the human species. At the same time that trick makes us belief that we experience fulfilment, not to be in the moment of fusion with the desired partner. The identification with somebody to be separated from that fulfilment is the source of that trick, that at the same time is the origin for a lot of suffering in life.

The physical body itself, due to its ability to experience physical sensations, opens up the possibility to directly experience consciousness in form of the self. (Tolle 2004) By focussing the attention on a situation with some stress in it, a situation that is affecting us very much, the physical body itself could be a "gateway". There is always the possibility to focus on the physical aspect. By paying attention to what exactly is perceived in the physical body and by focusing directly on the centre of that sensation with all the awareness, this very sensation will most possible fade and disappear. And very likely there is something else showing up that we are now focusing on immediately and directly. So sooner or later there will be no more sensations showing up and empty open space will remain behind. That empty open space is without any boundaries and it is limitless itself. It is consciousness itself, which is being experienced directly.

It is absolutely necessary to be true to what is perceived and to not mix it up with the mental or emotional aspects. For sure it is not possible to experience despair (emotion) or confusion (mental thought) in the physical body.

In the direct experience people often tend to imagine or see a picture of themselves going into the experience instead of focusing the awareness on the sensation itself. This is actually no direct experience. On the contrary this is separation itself and the subconscious attempts to continuously stay outside the process in order not to challenge the identification as an individual. In that sense direct experience means to experience the sensation fully by focusing on the total awareness of that sensation. This is like really being in the midst of that sensation and experiencing it from inside out, instead of picturing yourself going into that lake.

The unique gift of the direct experience is to give the opportunity to use every single aspect of life to experience consciousness directly and realize the illusion of separation. It is the realization of oneness and it is miraculously, so that this possibility is hidden just in a situation that seems to be so far away from that realization of oneness. The physical body itself can be a gateway to realize consciousness.

Experiment for direct experience no 7: (Tolle 2001)

Focus your awareness to the inner sensations of the physical body. If you are not able to sense that inner body you can start to feel your hands. Hold your hands in front of yourself towards your face. Be sure that your hands do not touch anything or rest on something. Maybe you want your elbow to rest on your knee or on a pillow. Close your eyes and ask yourself: "How do I know that I have hands?"

Feel your awareness going into your hands. There might be a tingle, heat, numbness or vitality that you feel. Whatever you call or feel it, it is life force. Feeling that life force is the connection to silence.

Now, let your awareness move to other parts of your physical body (arms, legs, feet, hips, torso, neck and head) until you can fell all parts of your body. Note that you can feel your whole physical body as one energy-field and bring your attention to that energy field for the next 5 – 10 minutes and be delighted with your meditation.

- This is what you are -

From story to structure – how avoidance continuously causes suffering

It is the avoidance of direct experience that continuously causes suffering and keeps humans in their own trance (of me and my life), that they describe by their name. In everyday life it is also often expected to share the individual trance in order to feel connected. In that sense the following story could be a good role model vicariously for many other stories in everyday life.

A woman is telling a good friend of hers about a conflict that she has with her son-in-law, who is such a bad person. She now has to take care of her daughter. When that woman's friend signalled to her that she does not know that person personally and cannot comment on him or on what problems he caused her. This caused a conflict between the woman and her friend, because she did not feel supported. The conflict grew when the friend finally told her, that she obviously has a problem herself with her divorced husband, which has not been processed at all. All this took place four years ago and both persons have absolutely no contact since then, although they were meeting regularly before.

This little episode is a typical example of how we continuously cause pain and suffering in our life and in the life of the people, we are connected to. There is the expectation towards our friends that they support our story - the way we put ourselves into trance – the story that we are so captured in. Friends are expected to condemn the enemy and consider us to be victims. A typical friend is somebody following us into our trance and strengthening this trance. When he/she feels sorry for us or supports us we feel better in that trance. All players in that game are losing because they go deeper and deeper into trance. Ponder for a while, which interactions are following in general and in that story. Now there is a problem with the son-in-law and a new problem with the friend. The positive intention of the woman was to feel free but she was gathering more frustration. The attachment to the story and the avoidance to experience the feelings underneath the story lead to an amazing amount of suffering. It is a high price to be payed. Instead of direct experiencing consciousness there is pain and suffering experienced in repeating and continuing the story.

Actually it does not matter how we react to a story like that, whether there is a moving against, towards or away, whether the emotions are expressed or suppressed. The reaction itself is a sign of trance already, characterized through the distance of direct experience caused by judgement and justification.

Direct experience is a medicine for trances like this. The word medicament comes from latin "medica mente" and means: "medical mind". As the problem is created by the mind there is the possibility to unveil that mind by looking into the inner structure of the trance. There is the possibility to take away the attention from the story towards the structure. In turning the focus of awareness to what is really perceived in the physical and emotional body and in directly experiencing that, there will be the self showing up automatically. The self is always there waiting to be unveiled.

By directing attention to what is driving the story this will be a gateway to the fullness of consciousness – a gateway to the self.

In this little episode above there are emotions (of fear, sadness or rage) and physical sensations that set the story in motion. As long as these emotions are not inquired and brought to awareness so that they can be experienced directly, the suffering and continued repetition of that story will continue. Subconsciously the woman was aware of these emotions, but nevertheless she continued to be occupied with the story instead of turning the awareness towards what is driving her. Unless it is the suffering that the woman gets satisfaction from, there will be no real solution or relief in changing that story anyway.

By projecting her feelings and the emotions onto her son-in-law, her unveiled emotion becomes an outside problem in regard to the relation to the world. Not even the case that the problem could be solved outside will touched the source of her identification as a separate individual. Instead there will be another possibility missed to unveil the trance of a separate identity. The trance (of identification) becomes stronger and stronger.

In contrary direct experience (of emotions) unveils every trance (of misidentification) and opens the door to realize consciousness itself. Directly experienced the realization of oneness opens and deepens – the self – experienced as empty space, silence and bliss. The only precondition to that is the willingness to experience directly what you are aware of physically, emotionally and mentally.

The little episode is demonstrating a conflict between the woman and her friend although there was no intention at all. Because she was not willing to support her story, the woman was missing the evidence of friendship itself. Therefore at this point the relationship ended.

If it is the direct experience of consciousness that we are longing for, stopping to support the stories that we hear from outside as well as repeating and believing our own stories, is the open door to that realization. If there is support of what is not touched by circumstances/environment and does not show up as a trance (and disappears again) there is the possibility (for everybody) to directly experience consciousness in many everyday life circumstances more and more.

Jaxon-Bear (2003) called this "A true friend" and this is friendship on a much deeper level.

Experiment for direct experience, no 8:

Go back in your memory to a situation that has some unsolved problems or triggers, some suffering and write down some keywords about it on a piece of paper in the first column of three.

In the second column you write down all the sensations in your physical body and emotions that you are aware of if you imagine yourself being in that situation. (These are what we (subconsciously) are aware of).

The third column is reserved for writing down all the reactions that you are aware of concerning the sensations and emotions of column 2 (These reactions are what we are identified with in our life – the root of our identification).

Now focus your attention to one of the sensations or emotions you wrote down in column 2.

Put aside the connecting story for a moment experience directly that sensation/emotion by focussing totally into the centre of the sensation/emotion – as if you are jumping into a lake that consists of that sensation/emotion. Focus directly onto that centre. Be that sensation/emotion. What are you aware of?

Focus your whole attention on the centre of all sensations arising until nothing more is showing up!

What is it that you are aware of? Who is it, being aware? Can that possibly disappear? Are you separated from what you experience directly?

> Is there anything needed to be added to this moment? Does this direct experience need anything?
>
> What if you go back now to the original story or the situation, what are you aware of? What if you would live your life from here?

Case examples

The following case examples are just attached at this place without any censorship or correction. In order to protect the privacy of the people and with respect to their integrity the name is just shortened. Otherwise theses case examples show in a typical way how inquiring takes place and how direct experience lead to the experience of consciousness itself in form of the self.

The first example is called: "Supernova of expanding joy" and it shows clearly the moving against tendency which is arising from the physical body while the second example "doubting can be a signal" shows very impressively the reactive tendency of the moving away arising from the mental body.

SUPERNOVA OF EXPANDING JOY

D. introduced her problem in the way that she did not feel happy although she was very successful in life. She avoided contact with other people because she felt great distrust to everybody. As she is very sensitive of human interactions, it is obvious for her when people are not authentic. Because she perceived the weak point in everybody (and let them know) and did not hide her strong energy while expressing that, she is facing big problems with other people. Therefore she withdraws from contacting other people if that is possible which does not make her feel happy at all. She came for consulting because she was longing for happiness.

K. Shall we explore that? How do you feel?

D. I feel so empty inside, somehow hollow. I feel like being already dead somehow.

K. How do you react to all that feelings of emptiness and hollowness, which feels just like being dead?

D. There is some condemnation in me. I do not want to have these feelings. I have enough and just want to be left alone and be happy.

K. What emotions are emerging with that reaction?

D. There is so much rage. I do not like it and want to get rid of it.

K. What is it that you really want?

D. I want to be happy and feel joy in my life.

K. How would you realize that you are happy?

D. I would feel alive. I would feel my life energy flowing and that is a good feeling.

K. Do you feel your life energy now?

D. No, not now. But I have felt it before in other situations.

K. What situation is that?

D. No special situation

K. What if you imagine experiencing this joy by remembering that situation in this very moment?

D. That's not it. This is just imagination, I really want to experience happiness now.

K. What do you need to feel right now?

D. I feel angry, there is a lot of rage and I am frustrated feeling that.

K. Okay, so please allow yourself to fully experience that rage – now.

D. It makes me angry. There is so much injustice in life. I did my very best all the time and my parents have suppressed all my aliveness and creativity and....

K. So in this moment – now – what if we just put all stories aside for a moment. You can have it back later, but for now just let us concentrate on the rage that you are feeling. What if you are falling into the centre of that rage now?

D. I want to beat them. I want to beat it. It is too much....

K.   Focus your attention directly on the centre of this feeling…like diving into a lake… be inside there, totally! And what if you even take away the label "Rage" and put also this aside for a while? What are you aware of now?

D.   There is so much energy… it is pure energy…and it is expanding more and more.

K.   Yes, that's right! What if you welcome that energy, welcome this energy expanding and experience that energy directly. If you are in the centre of that energy, what are you aware of?

D.   It is so difficult to keep that energy in the container/body. There is too much energy.

K.   What, if this container is just an idea? When you also put aside the idea of a container that keeps the energy inside for a moment. What are you aware of now?

D.   It is like a supernova. It is like a supernova expanding itself continuously all the time into a boundless, endless space.

K.   And what is in your heart? What quality are you aware of in the heart just in this moment?

D.   Joy. It is joy. It is a supernova of expanding joy…

K.   What if you look from here to the situation with your parents. What if you welcome all your life from there – now. Is there anything that needs to be added or changed?

D.   No, it is just my life. It does not need anything and is absolutely perfect. There is absolutely no separation in that.

K.   So you can see your life in the light of that expanding joy. Is there anything else you are aware of now?

D.   I realized that point, where I cut myself off from my own life force. It has nothing to do with anybody else. There is actually nobody to be responsible in terms of guilt, for anything.

K.   Compared to the situation that you were experiencing when you showed up here, how big are you compared to that experience?

D.   Not even measurable – there is actually no past in the moment.

DOUBTING CAN BE A SIGNAL

A. visited me to get help because of the fear she was experiencing. She seemed to have lost orientation and was suffering from doubts that effected and defined her whole life. She was also looking for more confidence in life.

A.　I am so fearful. I have Nordic walking sticks which I use as a weapon. I was not that extremely fearful before, but since recently....

K.　So what happened in order for you to become so fearful.

A.　Nothing really. I do not know. At least I am not aware of anything that happened to cause this tremendous fear.

K.　What is it that you really want?

A.　What I really want to know is how it is working to go deeper. But I do not go deeper, so I do not know.

K.　What if you do not know? What if you do not know how it is to go deeper? What is it that you are experiencing now at this moment?

A.　It is like fog and I cannot see through it. There is so much fear and so much doubt. I doubt so much!

K.　What is doubt giving you? What do you hope to get? Is there any positive intention in doubting?

A.　I want so be safe. By doubting I hope this helps to make right decisions.

K.　Is that what you achieve by doubting trustworthy in any way?

A.　No, not at all. Everything is veiled.

K.　How exactly? How does is this veiling happen?

A.　I do not know I am so unsure about that. So I think I have to go deeper to find an answer.

K.　What do you hope to find by going deeper?

A.　When I go deeper then the doubt will hopefully disappear automatically. On the other hand doubts keep me at the surface. I am afraid of looking deeper. Maybe if I go too deep this could have unknown consequences.

K.   What do you mean by too deep?

A.   Something could happen, which is unexpected and...

K.   So what is it that you are aware of at the moment when you imagine going deeper?

A.   I better stay at the surface. It is safer. I am afraid to touch a deep ground.

K.   How do you know when you touch deep ground? What are you expecting to find there?

A.   There would be a good feeling. But I do not dare to look deeper in case there will also bad feelings. I am so unsure and afraid.

K.   What bad feelings do you imagine waiting?

A.   I do not know. It was always very unclear and uncertain.

K.   What is it that you really want?

A.   I want to realize the essence of things in the depth.

K.   If you could see the essence of things, what would that give you?

A.   Oh, at the end that will give me satisfaction and power

K.   How do you know, that you have it? What would that give you? What would you be aware of then?

A.   I would be aware of the wideness. There would be no thoughts and there would be authenticity.

K.   What if you are aware of the wideness? There would be no thoughts and there is authenticity. What would that give you?

A.   Then I would be disappearing

At this moment the whole body relaxed

K.   What is there in that moment when you are not there?

A.   There is power and joy. It is unlimited and there are no boundaries.

Suddenly A. winces almost imperceptibly.

K.   What happened?

A.   It was like a beat. Like a shock, that suddenly brought me back here again.

K.   How does that happen exactly?

A.   It is like being thrown out of the self and suddenly the wideness fades away and I am back again. There is fear coming back too.

K.   What happened in order to bring you back again?

A.   Something came into my space, something from outside made me showing up again.

K.   What is coming from outside? How does the "I" suddenly arise?

A.   It is like a black dark something coming from the right side.

K.   How exactly does this black dark something come from the right side exactly?

A.   It is like a grey, dark, big, cold, polished stone that suddenly appears.

K.   Let us make an experiment. What if you totally focus on that stone for a short moment? What are you aware of? What happens?

A.   There is no fear and it is just wonderful and miraculous.

K.   In that moment, what if the stone shows up again and you do not move? What is happening? What are you aware of?

A.   It is like a message delivered by the stone. A message to stay true and experience the truth of what is.

K.   What are you aware of now?

A.   It is unlimited something,

K.   Yes it is unlimited. What if you could travel back to a time when doubts were showing up first time?

A.   Terror comes up,

K.   What if you welcome this terror now?

A.   I am inside wideness, the stone is also there. There is this ocean of light, I am diving in and out.

K.   What if everything goes into this ocean?

A.   Everything dissolves.

K.   How does this feel?

A.   It is like diving in – diving out.

K.   What happens if you dive out?

A.   Separation, form is coming back.

K.   What is this form made out of?

A. It is solid with a little spot of light inside.

K. If you go into that spot?

A. There is widening instantly. Everything arises out of that wide infinity. It is like the ocean.

K. Is there anything else? Could you be thrown out of that?

A. No, I have to laugh. It is that thankfulness. How could I possibly be fooled by doubts again and again?

K. This is the divine play, so that you can laugh again.

K. Where is doubt now?

A. It is not there.

K. So doubt can be a signal, thank you so much.

A. Why? I did not do anything.

K. For that I am thankful!

A. It is overwhelming.

REFERENCES

Almaas, A.H. 2004. Facetten der Einheit, Kamphausen Verlag

Bhagavad-gita. 1955. Reclam Verlag.

Balsekar, R. 2007. Wo nichts ist, kann auch nichts fehlen. Lotus Verlag.

Balsekar, R. 2009. Kein Weg kein Ziel nur Einheit, Lotos Verlag.

Castaneda, C. 1971. Eine Andere Wirklichkeit, Fischer Verlag.

Castaneda, C. 1995. Die Kraft der Stille, Fischer Verlag.

Gangaji. 2006. Der Diamant in deiner Tasche, Goldmann Arkana Verlag

Gribbin, J.2009. Quantenphysik und Wirklichkeit, Piper Verlag

Harding, D. 1988. Das Buch von Leben und Tod, Context Verlag.

Jaxon-Bear, E. 2001. The enneagram of liberation, Leela Foundation.

Katie, B. 2002. *Lieben was ist.* Goldmann Arkana Verlag

Maharshi, R. 2006. Gespräche des Weisen. Lotos Verlag.

Patanjali, 1976. Die Wurzeln des Yoga Otto Wilhelm Barth Verlag.

Pine, Red. 2006. The Platform Sutra: The Zen Teaching of Hui-Neng. Counterpoint.

Poonja, H. W. L. 1993. Wach auf, du bist frei. Kamphausen Verlag.

Stüben, K. 2010. Du bist frei und kannst das Jetzt direkt erleben. Windpferd Verlag.

Taylor, J.B. 2009. My stroke of insight. Hodder and Stoughton.

Tolle, E. 2000. Jetzt! Die Kraft der Gegenwart, Kamphausen Verlag.

Tolle, E. 2001. Leben im Jetzt, Goldmann Arkana.

Tolle, E. 2004 Torwege zum Jetzt, Arkana Audio.

Wei Wu Wei, 1982. Open secret. Hong Kong University Press.

Wheeler, J.A. 1974. The unviverse as home for man. American scientist 62

SIGN OF A COSMIC CODE TO BE VERIFIED:

Finding of a big chakra over Värmland, involving the Cheops pyramid of Giza. Built on the Fibonacci series of holy numbers.

A CIRCLE QUADRATURE IN THE VERY SOIL MATTER OF THE GLOBE.

G Anita K Westlund

WHAT DOES THIS VERY remarkable figures mean, and why, if not as an offered blessing to a world in need for renewing.

Chakra is believed to be about managing a transformational process in between human and cosmos, handing energies outside the mere view of today.

The very presence of this marvelous figure seems to be a call from the highest Consciousness, the quantum Lord. Involving natural constants as the speed of light, sounds, the globe sloping angle, as well as the time system of this earth and something called chakras (from Arne Groth compendium).

And, 377 times zeros is the distance found to the upper cross of our

Scandinavian romb, as well as to Stonehenge and Santiago de Compostella.

A SHORT DESCRIPTION

It has gone twenty years since I first observed that the churches of ours are very specially placed in the geography. At about one hundred straight lines of three or four hundred kilometers. Sometimes from sea to sea, or at least following our big valley systems. All situated at very special geographic places. Exactly including water falls, mountain crossings and metal bodies. How and why this multi function in line?

Furthermore the systems seem very old cause many of the places are from since long dried out lake systems,.. our region is still rising four meters a thousand years from the Ice. As well as the lines to some level seem to be thematically intercoupled.

Those seven Värmlandian church crosses I've studied are involving about sex hundred points, in seven-eight former or present valleys, from Västerdalälven to Glomma. The two biggest systems, Glavsfjorden and Fryken, are also clearly involved in the middle age culture south from the large lake Vänern. Telling about the same sensitivity since ages. The system seem to have been known for at least ten thousand years of land rising variables.

Ancient people obviously had a healthier sense of cosmic reality which modern citizens seem to have lost.

Sad, cause our mental knowledge is superior only if we have trust enough for combining our natural given intuition! People of today seem to be more and more devoted to superficial esteems, seriously threatening the earth.

These manifestations are far more than random requisite. Telling about deep layers of system information.

Five years ago another artistically eye seemed to has been opened in me. I had in vain tried to have help from the university, at least with the computer checking. But after some months with the first glance of Arne Groth's compendium of dowsing mathematics, for me absolutely new, and Dan Mattsson's book of stone mathematic from former cultures south from Vänern, I started to observe an overlaying geometrical pattern and these ever repeated numbers of Fibonacci.

The material has been processed through five years, and this latest month when formulating this very chapter, I'm happy to tell more of fundamental data but in the sides below: Except the exact 12,0 periodicity in between the two centers of Gräsmark and Giza, I see the 354 Swedish mile distance is talking about the holy ratio of 3-5-8,(4 instead of 8 just means a lower octave). It really represent the well known Pythagorean axiom, still more pointing a spiritual culture connection between the two centra's.

The real distance of 354 was early registered a tempered f, as the same in scale, 236 millimeter, was tempered b. The very sound of our Värmlandian songs, as well as the tune of the bottom lines of Cheops (231m) and our Scandinavian romb sides (235km).

Furthermore observed the height catered of our romb (185 km) have the same digits as the wave length of chakra here at the 60th degree latitude, 1,85 meter. First found in the bottom of Cheops, 1,85 x 125, as well as at another 60 degree place in Sweden... Still more proves from a universe telling realities. Telling about a great Consciousness behind our mere understandings, and of which we supposing are dependent.

How to use this information, and why now? For me it's an answer to the need of deeper integration in this time of nonchalance. Needing insight of sorts not yet accepted in present society.

THE VERY FINDING

1. First coin

A mediumistic friend of mine twenty years ago told me I had a rare and strong energy point at the top of my mountain. And she told me to go there singing!! Some later she came back to listen to my vioce in the old church of mine, Östra Ämtervik. I really had questions, but she was in a great hurry for visiting another church far from here before going home again.

In pure frustration I grabbed a linear and connected the two points at the map. Became exalted when seeing the line passed exactly through Tanum at the Atlantic, the most known ancient reminiscences of this country! I just had to investigate.

The line was exactly passing Vårvik, Stavnäs, ÖstraÄmtervik, GustavAdolf and Insjön. All with the same sort of position in the landscape. And far north crossing Iggesund at the old mouth let of the Ljusnan River into the Boltic. A six hundred kilometer straight line in between two seas. Connecting old culture places in sex river valleys, all running from the high mountain district of Sör Tröndelag south from Trondheim. Earlier weekly visited since the eighties by ufo´s, of which I knew nothing until ten years into my project.... Validity comes first!

2. Verifying this first point

More lines through my church?

The first scanning gave twelve lines with about eight churches at each. With hundreds of kilometers and just one single millimeter of deviation, one kilometer, confirming this no random perspective. One line was exactly passing through Vismum Kil and Södra Råda, both seen to be very old, and straight on to Vadstena, the monastery of santa Birgitta, and except Fryken the other deep rift valley of this region.

The other end of the line passed Åsnes/Flisa, center of the 1500 years old Solör Country crossing the mountain sides of Norway and Sweden. From the fiord of Oslo to river Klarälven dividing the delta of our Värmlandan acreage.

A second check of lines showed on double the amount of lines and points if regarding the peripheral points more than the central, a map problem cause of the globe. But special check towards google map showed acceptable deviation at large, more at the diagonals and hardly nothing at the north–south ones. I´ve counted the less number of points.

3. More crosses?

The seven crosses presented has a large numbers of churches, around 50-100, sometimes interrelated and in team. Except Gillberga and Högerud in the Glavsfjorden/old Glomma system being peripheries from large crosses of the Västergötalandian medieval landscape, the map often point to perhaps ten thousand years of culture coming from west, from an open salt ·Atlantic. When still ice over the inner areas of today.

Other crosses were Sunne at lake Fryken, Nedre Ullerud at the Klarälven river, By/Säffle where the Glavsfjorden system met lake Vänern, and Eda just at the border of the mountain heights of Glavsfjorden, a former continuation of the Atlantic connected Glomma river before the land rising turned it around at Kongsvinger. Obvious is also that the Glomma once divided southwardly from Eda, through the Järnskog-Silbodal- Silen-Lelången system out into Vänern, as well as the still older Foxen-StoraLee system directly went back into the Atlantic through Foss/Munkedal and the Gullmarsfjorden of Uddevalla.

The two crosses of By and Östra Ämtervik are about the same size and far more largest, about 120 points with one third south from lake Vänern.

4. Taberg, another special and important information

First with the drawings through By ten years after starting, the meeting cross of the Glavsfjorden and Fryksdalen lines at Taberg, a rare mountain horst south from deep lake Vättern to whom our own Tossebergsklätten at the same line is parallel, my eyes got still more opened for underlying electromagnetic ideas, mental or not.

5. 24, connecting the earth movements? The importance of Östa Ämtervik as a first point of reference

The distance ÖÄ–Taberg is 23,5 Swedish miles (235 km) which was associated to 24hours. My interest grew still firmer when even the northern meeting point between the crosses was exactly the same, 23.5, and with the same deviation. This time it was the special sanctuary line from Vadstena-ÖÄ meeting the prolonged Taberg-By.

6. Involving mathematics of ancient stone patterns

But it was with Dan Mattson's book in mathematic of ancient stone reminicences, I got a clue of larger overlying patterns possibly inherent. Before it was just church lines.

7. Overlaying patterns of Varteig, and still more of 235 kilometers

Drawing all the seven church crosses at the same paper, a new builded point in Varteig at Glomma showed up as outstanding. Having contact with most of the others. Why?

So, as the distance Hedmark to Varteig also showed to be 235 kilometers, I could hardly leave the papers. And recognizing the distance from Varteig to my ÖÄ beeing half of it, really implemented the importance of this project starting in my church of ÖÄ. Furthermore the distance from Varteig to Vadstena was about the same, but yet not exact...here was more to note.

8. Trondheim-Rom and still another perspective

When the line of Varteig-Hedmark showed to involve Trondheim in the north as Rom in the other end, new perspectives begun to emerge... Was it some deep esoteric on the go??

9. A well known ley line holding an important cross at Insjön

Varteig for sure was found at our locally very well known ley line of Ekshärad–Lysvik–Gräsmark, which in itself seemed to lack a large church cross. But now building another edging one in meeting the first line from Tanum–ÖÄ at Insjön in the Siljan/Dalälven water system.

10 A first Pythagorean hint and necessary stability reasoning. The short catheter is 146 kilometer! Noticed a straight angle perhaps was found in Gräsmark, and the upper east triangle reminded a Pythagorean one with its ratio of 3-4-5. Deviation was three percent but worthy to go on with as the distance Gräsmark-Insjön had no dependency to the sea level, like the Varteig following the land rising of four meters a thousand years for sure had. Another factor of in secureness was the northern sharp angle in Hedmark. But anyway the Gräsmark cross started my association towards ancient Greece,.. and Egypt!

11. The Petrus Cliff of Roma exactly at the pre lengthened line from Trondheim-Varteig. And Flisa- Gräsmark-ÖÄ passing the Nile delta!

When the truthfulness of Trondheim–Hedmark–Varteig–Tanum–Rom was confirmed, my interest fall at the other long line from the Hedmark cross, the one from Flisa to Hov south from Vadstena. A first inquiring notation was the passing of the Nile delta of Egypt. Saw the height from Rome towards it seemed to be about half the Hedmark distance,.. a fully trine?? As fond of drawings this was exciting, but still I didn´t associated the Cheops.

12. The Cheops and a double trine making a perfect romb over Europe, connecting two spiritual centras?

Suddenly observed the distance from Rom to the Nile could be the same as the one to Hedmark. 142 millimeters from Roma exactly made a cross in Giza, the Cheops!!

13. Gorlovka, the fourth corner

Draw the double height from Roma and the pencil stopped at the only mountain top of the Rone delta, 369 meter, just behind Gorlovka in the Ukrania. That means the distance of Gorlovka-Giza could be the same as Giza-Rom. Which is! Another unusual trine and a complete 144 romb over Europe! Which means?

Much later: It's needed to observe the Ukranian war started in Donetsk in the very same area, as well as the mountain top holds the same Fibonacci number (370) as of the Scandinavian length diagonal and the Pyramid sides.

All this numbers and geometry is far from hazard.

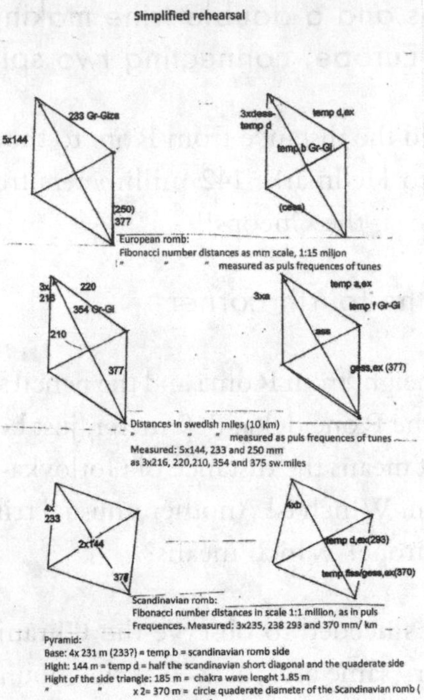

Fig. 1 G Anita K Westlund

MORE OF OBSERVATIONS

All 2013 I actively tried to melt the true challenging information of Arne Groths mathematics:

14. 144, started to associate with the Fibonacci
    numbers of Cheops. As Atlantis migration The map
    scale distances of 143 millimeters were double
    the exciting as they were just scales of a map.

Still showing the very digits, so what means this 1:15 million?
From Arne Groth the Fibonacci 144 is often found when in harmonious circumstances. And the pyramid numbers of 144, 233

and 370 (375,377), will repeating be found in all the geometry patterns, European as Scandinavian. Supposing positive,... as holy?

15. Gaia

12 is seen as a holy number, and 144 is twelve times itself and at the twelfth place in the series, and the only whole quadrate up to the length of fourteen digits. Groth mentioned the coordination in between Pi and the Fibonaccis is high enough first at 144, for constructing a circle quadrate as around the pyramid. Which seems to be coupled with the idealistic globe of Gaia as Plato talked about. A perfect Gaia is situated 23 kilometers under the poles and 34 under the equator.

Please observe Groth didn't knew about the Scandinavian romb when asking him. Now he's gone since fifteen years.

A rational thinking chemical engineer,.. imagine to digest these numbers!

I'm very grateful I already had met the intuition in me, as well as further studied it a bit even i others. The engineering behind maybe twelve thousand years of inner pyramid knowledge would otherwise be totally wrecked. Said to be a rest of a seventeen thousand years old migration to Egypt from an Atlantis in decline, perhaps to ascertain their universal knowledge of nature. They are said to have been good at geometrics and stone frequencies.. too good for their society level.

Their managing a pyramid is acceptable, but finding their ways to this manifestations in nature of the far north is more than challenging, if not for Atlantis's or followers coming here!

16. Ancient intuition, science and the superficial idioms of today

The wise ones of Atlantis seem to have admitted a higher degree of intuition than normally used in society of today. But time supposing goes forward, us growing into still deeper accepting and consciousness. Yet, their reminding about human resources is of really

263

great value. Seemingly about an overlying structure, wholeness, of which even the holography and quantum physics of today seem to be speaking.

David Bohm and Karl Pribram already in the early seventies begun to put words on old esoteric, not to talk about Einstein. Much has happened since than but the political society seems dangerously retarding. Our best researchers are since long talking about an unlimited consciousness, but where is the practice.

17. The second number 233, trusting and verification. A new southern cross of Hedmark With renewed eyes I started to honor my lines being no already known leylines. New stuff.

As earlier told I noticed the alikeness of my first Pythagorean triangle with the pyramid numbers. Recognized the 238 kilometer, 233 if yet with some deviation. But if a meaningfulness in the curtains, it for sure never would have acted imperfectly!!

So, further counting at the less sensible Gräsmark-Insjön showed a perfect cross of 235 kilometer in southern Hedmark. Twenty kilometer south from the first one which still had to be charged cause of the 375 through Europe to be respected.

18. And if one bowel of meaningfulness, there is two: A time process and a perfect romb. Why shouldn't the right angle cross of Gräsmark holding more than one holy triangle, at least one along this stable line! More cause deviation of a long derivate sharp angle line than some few thousand years of continental rising could have caused the 238 kilometer side shortening into the south Hedmark, as the point of Varteig being pushed to Slevik at the water line of today. All numbers of the triangle are now perfect. As well as the rombside down to a new Vadstena point become

exact 235 kilometers, to a supposed rift cross in between two water systems in lake Vättern outside Motala-Vadstena. Confirming the fourth romb side up to Insjön becoming 238 kilometers, not more than one percent deviation. A perfect romb relating the truth of Cheops. All the way round!

What about if this holy figure was blue printed before the pyramid was built! And partly detected.

19. The same diagonal numbers found as in the pyramid, 370. And not due to scales but to the very digits!

The south going diagonal 370 kilometer is identical with the one in the pyramid side (370/2 meter), the circle diagonal of a circle quadrature long combining heaven and earth, spirit and materia.

It seems to have been of most value to document this esoteric. Perhaps their feeling the irrational Pi-circle "couldn´t", but the firmer circle quadrate construction on holy numbers could! A knowledge demonstrated by innumerous of stone remains. **But why not parallel to the pyramid high itself, but to the side?**

20. Pythagorean quoting's

Pythagoras said the world is build on numbers. Perhaps we have been too ignoring!

Furthermore he claimed the soldier coming to kill him, not to rub his circles.

Perhaps of help when looking behind to change our obviously incomplete mattering.

21. 1:15 million?

What tell the map scale of one to fifteen millions, twice revealing numbers in the serie of Fibonacci?

This very autumn I came to hear a culture program at TV mentioning there seem to be a law telling doubling the size gives fifteen times more of growth!?... (uncertain, but at the numbers). Once more telling a hidden truth behind certain numbers, a truth free from mere materialistic inclinations. Just what our circle quadrature are to remind of?

22. The third number of 375, a constant telling about chakras

The presented 370 Scandinavian diagonal down to Giza is 3750 kilometer, only 2 and 0.5 percent deviation from Fibonacci´s 377. Please remember 375 seem to be a natural constant involving our speed of light divided by speed of chakra and 14400. (Arne Groth)

SOUNDS

Sound interconnections b, d, fiss, successively <u>moving from minor to major</u>: d, fiss/gess, a?
Again seemingly confirming a dynamic behind.

The map distance between the two centers is 236 millimeters. The same numbers as in the upper romb (235–238 kilometer) responding to the pulse frequency of low b (471/2=236), which is in the most frequent harmony of Värmlandian melodies (b, d and fiss).

A new build harmony can be seen as rising from the European romb sides into a major clinging of d, fiss and a!

D comes from the Scandinavian new diagonal of Slevik-Insjön, 293 kilometer, with a sound frequency of 294 as well as all the big romb sides of Europe are swinging the d. As the frequencies from the Fibonacci number of distances around 2160 kilometers.

Fiss is coming from the length diagonals of Scandinavia and the pyramid side, 370 kilometer or meter. Which is exactly the frequency of fiss (370), while the longer one of Europe, 375, exactly corresponds to gess, 377. Just 0.5 percent of deviation, but together with the 2% to the rombs this could be a sign of a hidden factor from not yet recognized chakra flew variations.

But no less exact enough to build figures telling about an ever dynamic universe of knowledge to be reverenced, whatever it is.

Furthermore 1,44 is the relative frequency of gess. Gess, as in the pulse frequencies of the new found full length European diagonal (377).

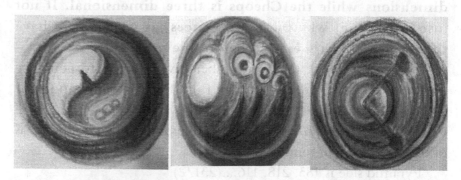

Fig. 2 G Anita K Westlund

Intuitive paintings from before project was started, all three from the same day in 2000, f- and g-clef and the angle of time.

THE HOLY SCALES IN MUSIC (ARNE GROTH)

The holy scale, 1.618, comes when dividing two following numbers of Fibonacci, higher the better.

The quote of single digit numbers gives c1, g, a, ass and c2, from where all the other twenty one pure quotes is derived. Lowest common denominator (Minsta gemensamma nämnare) for all this is 14400!

The holy numbers is found in the design of flowers, self refreshing water movements, cyclones and planetary systems as well as galaxes (see a special paper) A natural creating pattern we seem to have missed. Furthermore, the third number is the sum of the two previous.

MORE COMMENTS

A scale of holy numbers from a medieval monk projecting one of the most interesting ancient culture remedies of the world! Pointing to a vibrant and dynamic consciousness behind mere materialistic references.

Our Scandinavian figure is documented in just two dimensions while the Cheops is three dimensional. If not just to be seen, why these differences, as we seem to have the orginal? Nature existed before the pyramid!

And the trines of Cheops numbers over Europe to point the northern out, at 60 degree latitude! Both built on **the important spleen chakra waves, 1,85 meter**.

185 times two is 370, the circle diameter of the pyramid side as of the Scandinavian romb.

Pyramid side is 185, 218, 116....(231/2)

European trine is 187, 217, 110..(220/2)

Scandinavian romb is 185, 235, 146...(293/2)

Pyramid snit is 185, 116, 144

Have I mentioned the idealistic Gaia waves is thought to be most available here at 60 degree latitude, along this open sea border outside the ice, where the globe surface could be expected to be the most thin. That is, more easily transforming cosmic waves from a universe (in constant change)? Some people says the globe is written inside an isocaeder, meaning twenty centers like Giza. But it was here there was a warm Golf stream and land enough for a culture to manifest.

– May we collect from the threat of zoombies mentioned below before it´s too late!!

ENDING

The last two years I more and more have been involved in what seems to be an alien story trying to wake us up. Specially the children is very exposed until we´ll find our way out.

– If claiming a society in need of three globes, help is really needed!!

The problem is more than mere materialistic. We seem to be missing in deeper connections of life. Perhaps of ethernal dimensions respecting each and all, involving timelessness and reincarnation, consciousness and life beyond this special materialistic concept. Holography and quantum physics even in us to be trusted.

If in me, we all have the truth of wholeness/oneness inherent, just to let it out renewing the world. It´s a constitution possibility. But we for sure have more or less easy to manage it. May the God bless us!

HOLY SCALES IN OUR SUN SYSTEM

Distance in astrologic units (AU), from Sun to Earth 149,6 milj. km = 1

Planet	Distance AU	Distance Ratio F=0.33	F=1.7	F=6	F=10	Mass %
Sun						99,85
Mercury	0,3–0,5	1				0,00002
Venus	0,7	2				0,0002
Tellus	1	3				0,0003
March	1,4–1,7	5	1			0,00003
Jupiter	5–5,5		3	1		0,1
Saturn	9–10,1		5	(2)	1	0,03
Uranus	18,3–20,1			3	2	0,004
Neptune	29,8–30,4			5	3	0,005
Pluto	29,7–49,5			8	5	0,0000007

The distances ratio too is building series of Holy Scales, if not counting the tenth potentials:

1,7	1	ca1
0,33	2	-
6	ca 3	3
10	-	5

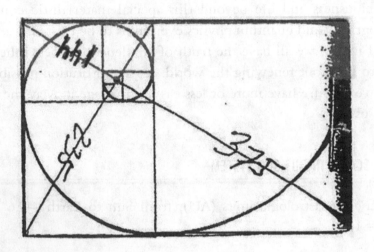

Fig. 3 G Anita K Westlund

Satellite picture of an hurricane clearly show the principle of the Holy Scales, with it's rectangle diagonals as follow: **89,144, 233, 377**

BIBLIOGRAPHY

1) Groth, A. Slagrutan avslöjar naturens matematiska hemligheter (Dowsing reweals secret mathematic of nature), Ekshärad: O–Liv, 1998.

2) Mattson, D. Jordstrålning hälsa och forntida vetande (Earth radiation, health and ancient knowledge), Nyköping: Nyköpings Tvärvetenskapliga bokförening, 1991.

3) Wilber, K. Det holografiska paradigmet och andra paradoxer (The holographic paradigm and other paradoxes), Göteborg: Korpen, 1986.

MUSIC AND CONSCIOUSNESS

Alexander J. Graur, Ph.D.

University of Torino, Italy Post-graduate School in Health
Psychology Member, The New York Academy of Sciences
info@medicamus.com www.medicamus.
com You Tube: alexander graur

MUSIC: DEFINITIONS

MUSIC IS A COMPLEX phenomenon; as such it doesn't support a unique and comprehensive definition. The various definitions gave in time concerned various aspects of the phenomenon, from the inner structure to the effects on the psyche of the listeners, considering all the aspects involved, from compositional techniques to the emotional and social implications of the music.

The most common definition disseminated in today's world–and the most approximate and misleading, is illustrated by the sentence; "Music is an Art", period. It appeared toward the end of the eighteenth century and the beginning of the nineteenth with the philosophic and cultural stream called The Romanticism.

Romanticism intended the term "Art" as the result of a not-so-well defined concept called "inspiration" in composing the music; what we could call today the unconscious of the composer.

The romanticism considers the uniqueness of the composer's unconsciousness as the sole creative force in the art.

It is not totally wrong; it is just largely incomplete and deceptive.

Reality is, since the dawn of human civilization until the beginning of the nineteenth century the music was considered, taught and used as a science. It was part of what the Ancient Romans called "quadrivium", the curriculum of studies which included, in order: Arithmetic; Geometry; Music and Astronomy. A logical order, one cannot understand and apply Geometry without knowing the Arithmetic, and so on. This logical curriculum of studies is common to all the cultures, Western or Eastern, because formed the students' minds to the logical thinking and its applications in the real world.

Music is an interdisciplinary science. It is governed by laws belonging to various disciplines as: physics, mathematics, physiology; to the specific laws of the phenomenon called Music and, above all, to the laws of what we call Logic, any type of logic.

The musical composition, as the creation of music is called, follows both the rules of the general musical architectures and those rules specific to various cultures and musical systems. The same laws applied in Western and Eastern, Northern or Southern cultures of the world, in exactly the same way the laws of physics or mathematics are universal. That is because these laws are created by, and deducted from, the reality itself; they are not invented ad hoc. The human beings have the same peculiarities all over the world; the human organism and its physiology are the same for all. The musical creation is a product of the human intelligence; in the same way the affirmation that one quantity plus one equal quantity gives two equal quantities, in equal conditions.

One plus one equals two even in the Universe, at least from the current point of view of the average human Logic. The Music is working with this type of laws.

That's why concepts such as "inspiration", "special talent" and alike are responsible of the final work in a very restricted sense. In order to be able to express his/her ideas in the musical language the composer must patronize the specific music compositional techniques.

These techniques are based on the laws of the interdisciplinary Science of Music.

In between the image of the composer who, caught by a sudden attack of inspiration, looks to the moon shinning in the sky, sits down at the piano and plays the whole "Moonlight Sonata" as some Hollywood C movies used to show us, and the reality of the final score of this Sonata (opus 27 number 2) there are at least various month of hard working. During all this time the composer is applying the laws and rules of the Science of Music, as strict and mandatory as any other laws of science. Another very important aspect, worth to be mentioned: the composer signed (Ludwig van Beethoven) at the end of the score of his fourteenth piano sonata, opus 27 number 2, in c sharp minor; period. The title "Moonlight Sonata" was given after the first public performance of the music, by a listener who said: "To me, this music reminds the moonlight reflexes on the placid water of the lake Konstanz". Which is a nice image, very commercial as the history demonstrates; a pity that it is also a very personal, subjective image; to another person the music will suggest a totally different image, or no image at all.

As a professional composer and a neurotherapist I give to the complex phenomenon called Music three definitions, regarding different aspects of it.

The first definition is: "Music is the science which organizes the sounds in logical structures". The second definition is: "Music is a language" (see further in this paper for details).

The third definition, more complex, is:" Music is a complex phenomenon which develops in five dimensions.

Three are space related (in a Cartesian system of coordinates, the X axis, horizontal, is the melody; the Y axis, vertical, the accompaniment; their product (X.Y), the Z axis, is the music as we received it).

The fourth dimension of Music is the Time; this dimension is intrinsically to Music, which exists only in time;

The fifth dimension, which provisory I define as "the aesthetic dimension" it is more complex still. It refers at the action of what I call "the Brain-Mind-Brain-Organism (BMBO) System".

Briefly: the brain receives the musical information; the mind processes this information and transmits the single elements which compose it to various zones and components of the brain; these parts transmit the single information, through the physiology of the nervous system to specific organs of the body (feed forward action). These organs react and transmit in turn the result of their actions to the brain (feedback action).

The process is continuous, until the entropy creates the stability of the whole system, in resulting of a new homeostasis based on the musical data and the reaction of the BMBO system to it.

The BMBO mechanic is different for the two categories involved in the process of composing/playing/singing and listening to the music. These two categories are: the producers (composer/interprets) and the consumers (non professional musician listeners). The specific and determinant aspects of each category and the relationships between them and the Music are the matter of this paper. (1)

PERCEPTION OF MUSIC

1. Perception in Space

Sounds are perceived both at a conscious and mainly at a subliminal level as organized information in five dimensions: spatial, temporal aesthetic, as mentioned above. **The Spatial Dimensions** are: two Horizontal: the Rhythm and the Melody; and one Vertical: the Polyphony in all its aspects (including the Heterophony and the Harmony).

The Temporal Dimension is the evolvement of the three Spatial Dimensions in time.

The Aesthetic Dimension is the conscious response at these elements. Music acts in time; the role of memory is to connect the spatial elements to the temporal dimension to obtain the psychological reaction. But this reaction is not only psychological; the body reacts at a physical level as well. The psycho- somatic chain initiated by the information- in this case the music- depends on the physiological and psychological factors specific to each person. The other reaction of the organism at this type of input is a somatic- psychic one. The presenting state of the physiological balance or unbalance of the organism determinates the decisions at the psychological level.

THE HORIZONTAL DIMENSION

The Rhythm: the Ancient Greeks considered rhythm as "the law ordering the movements" (Plato) or "the order in organizing the accents" (Aristoxenos).In the sixteenth century Glareanus defined the rhythm as "principium movens" (the organization of the movements).

The rhythm organizes, coordinates and subordinates the movement of the music's elements. In any musical rhythm there are three elements, organically connected: the rhythm itself, the metrics (or measure) and the speed (tempo).

Rhythm itself has infinities of forms as a result of the combinations and organization of the values – the positive values (sounds) and the negative values (pauses).

Metric (measure) is the frame in which the rhythm itself evolves; it is expressed by the periodical points of reference called bars, which determines the rhythm's evolution in time.

Speed (tempo) determines the amount of the rhythmical movement used in a determinate period of time. The rhythmical patterns are the movement itself; while the measurement units of this movement are the metrics.

Melody is a succession of musical notes (pitches) in time. In Music language the melody has the same meaning and structure as the sentence in the spoken language. It is the mean to express and transmit all – from pure ideas to personal feelings.

The parameters of Melody's structure are:

a) **The number of pitches**. According to this parameter, the Melody could be monochordal (one single note, repeated) or polychordal (two and/or more different sounds). The denominations comes from ancient Greek: (mono = one, single);(poly = more);(chordos = tone, sound).

b) **The quantity and quality of the intervals**. The distance between two different pitches in sequence, called interval, defines the whole expression and communicability of the melody. The most known organization of the sounds, the Temperate System, uses the step as basic value for the distance between two different pitches. This value is composed by nine commas (the comma is the smallest distance unit). The Tone is divided in two half steps. The half steps have different values and denominations; the diatonic half step has five commas, while the chromatic half step has four commas. Because of this measurement system the intervals (which are quantities) has also a quality, according to the type of half steps contained.

For example, the Third is a quantity (distance between three different sounds) and has a quality (four diatonic half-steps= major third; two diatonic half-steps + one chromatic half-step= minor third); etc. The alternation of pitches creates a line called Melodic Line.

c) The **directions** of the Melodic Line. There are three basic directions: linear, ascending and descending. The combination of those basic directions gives the characteristics of the Melody. (Example: linear+ ascending; ascending + descending + linear; etc.) This aspect of the melody is of a primary importance, especially for therapeutic needs. The possibilities of the Melodic Line in creating a really mind-body influential music are infinites.

d) The **number and order** of the Rhythmical Patterns. The Melodic Line could be organized in time by using one or more rhythmical patterns. When one pattern is used there is a mono rhythmical melody. When more patterns are used there is a poly rhythmical melody.

THE VERTICAL POTENTIALITY

The melody has within itself a horizontal trend as well as a vertical potentiality. The sounds are generated and generate. Any sound belongs to the row of harmonics of other sound; and any sound generates other harmonics (sounds).The higher the sound, the more possibilities to belong to more harmonic rows, generated by different sounds.

For example the C5 is: the 8th harmonic of C1/ the 12th harmonic of F1/ the 11th harmonic of G1/the 7th harmonic of D1 etc. In turn it generates the superior part of the harmonic row of those sounds – and 2 or 3 harmonics in the inferior part of the row. The possibility of a high sound to generate lower sounds is determined by the nature of the resonant that produces it.

The relations between the sounds it contains give the Vertical potentiality of a Melodic Line. If the sounds belong to the first eight harmonics of the same generator the vertical potentiality is relatively poor.

The variety of Verticality increase when the sounds of the Melodic Line could be considered as belonging to the harmonic rows of more generators.

It is the difference between monotonality, polytonality and free harmony that permits the Vertical Dimension of the music to be easily perceived as a sensed dimension.

THE VERTICAL DIMENSION

The Melodic Line (Rhythm and Melody), when emitted simultaneously by two or more resonators produces two or more layers of sound. Those layers create the Vertical Dimension of the Music. There are four categories in organizing the layers: the monophony, the heterophony, the harmony and the polyphony.

Monophony : is the organization in one layer, also called unaccompanied monody. Similar to the melody, the difference is that a monophony is produced by different voices and timbres playing the same melody at unison. Monophony is the main technique in playing ritual music since the dawn of the humanity. Singing and playing the same melody, with the same rhythmic patterns and the same metrics gives the sense of collectivity. It strengthens the feeling of being part of a bigger community; the individual is considered as a "zoon politikon" – a social human being.

Heterophony: at least two layers having independent directions, coincidentally joining at random times. The joining moments are incidental, not pre-determined. This is another ancient and basic way of practicing mainly ritual related music. It represents the reality of an individual who has a life of his own, marked by random moments of socialization. Rediscovered by the modern music (Ligetti, Penderecky, Foss).

Harmony: the organization of the sound in at least three layers, produced simultaneously.

In the Temperate system those layers are distanced by at least a Third. The rule of organizing harmony is to consider it as a function of the melody (the vertical aspects of the horizontal dimension).

In this way the main layer (the Melody) creates complementary layers underneath and/or above it. The sensation of space is further amplified by the use of various timbres, one or more for each layer.

Polyphony: the simultaneous structuring of several layers of sounds, each of them formed by independent melodic lines. In the Temperate system each layer is built according to the Horizontal Dimension's rules of rhythm, metrics, melody and the vertical potentiality. Each simultaneous aspect is controlled by the Vertical Dimension's rules of Harmony and Counterpoint. In this way, there are two spins, one horizontal and other vertical, acting together in the same time. Due to its strict and complex rules the Polyphony represents the highest form of organization of the sounds in space.

2. Perception in Time

Music is basically a phenomenon that develops itself in time, more than any other forms of Art and exactly like human thinking. Time has two aspects, as we perceive it: the real and the psychological aspect.

Real time, also called ontological time, is organized in hours, minutes, seconds, etc. Psychological time is the way each of us perceives and interprets the events of real time.

The Temporal Dimension of music is the way in which the components of Spatial Dimension evolve in time, both at the real and psychological levels.

There are multiple relations between: the real duration of a musical piece, the technical means of composing music and the auditor's sensation of time passing.

These relations can be classified in two categories: the first category includes music that follows strictly the real time. It creates the sensation of peacefulness and relaxation. The auditor (or patient) feels himself sympathetic with the music, and opens his mind to the information the music transmits.

It is the moment of ISO or Similia Similibus Curentur (the similar cures the similar). The second category is composed by music written in contrast to real time dimension. It delays or anticipates the ontological time. That is the principle of Contraria Contrariis Curantur (the opposites cures the opposites).

Example: a musical piece which takes five minutes to be played in real time is perceived as longer or shorter by different listeners in psychological time, according to their personalities and the physical and psychological presenting state of being.(1)

CONSCIOUSNESS: DEFINITIONS

Like Music, Consciousness is a complex phenomenon. Let's consider a perfectly cut diamond. How could we define it? Is it a stone? Certainly. Is it made by carbon? Mostly. Does it have any practical value? Tell me....

The evaluation depends on where you look, what you are looking for, what are you looking to, under what type of light (artificial or natural), why are you looking for, what you intend to do with it, and so on.

I do not believe we will ever have a comprehensive definition of the consciousness. But the quest itself is a proof that consciousness exists.

As a composer and neurotherapist, mainly regarding the music-based neurotherapy, I started from the celebrated Descartes' syllogism: "Dubito, ergo cogito. Cogito, ergo sum" (I doubt, therefore I am thinking; I think, therefore I am). The consciousness as the result of the brain, mind and physical activity expressed on the psyche and behavior, single and collective, with the prevalent role of the mechanics of memory in storing and using the information.

Briefly, some aspects of the phenomenon considered by various disciplines:

- The consciousness as self-awareness
- The phenomenological consciousness
- The introspective consciousness
- The moral consciousness
- The self-consciousness

As you can see, so many aspects of a complex phenomenon (yes, it is a really huge diamond with so many facets). We can also consider that the moral consciousness, the self consciousness and the free choice are the results of a historical and cultural development of the human race. Music is implied in all these aspects, in various forms and ways, because it is a final product of the human mind, implying the role and mechanics of the brain, mind, memory, behavior, anatomy and physiology, physics, mathematics and logic; which in a way or other are all various facets of this diamond called consciousness.

I have to mention at this point the very important and effective initiative of Stuart Hameroff, David Chalmers and the University of Arizona's Center for Consciousness Studies, which organizes every year the interdisciplinary international conference "Toward a Science of Consciousness" exactly with the goal of trying to give an as much as a comprehensive definition of the phenomenon while presenting the many facets of it. (website: www.consciousness.arizona.edu).

MUSIC AND CONSCIOUSNESS: EXPERIENTIAL ASPECTS

1. Out of Body Experiences

The out-of-body experiences (OBE) are a reality frequently present in psychiatry. Two types of OBE are considered: as a consequence of a strong psycho-physical trauma or as an effect of some drugs, natural or artificial (the other type of OBE, as a result of meditation in a spiritual, religious environment, it is not classified as OBE in psychiatry).

For example, in the PTSD (post-traumatic stress disorder), the tell-tale of the OBE is always present; all the victims of rape tells about these kind of experiences; themselves as a victim of the physical act with all the feelings and sensations; also themselves looking at the act of rape as themselves as victims, from somewhere above in space, as an "alter ego" (another myself), with the feeling of impotence to react to the violence while feeling it physically and psychologically.

Anyway, the OBE are a clear proof of the existence of the consciousness, because demonstrates the capacity of the human mind to act simultaneously on different space and time levels, in meantime keeping the ties between the various aspects of the split reality, objective and subjective.(2)

In this paper I will present another type of OBE, seldom described; it regards the Music.

An orchestra conductor (and at various levels all the professional musicians, of all styles of music; rock, jazz and pop enclosed) are doing the following operations in the same time:

1) Follows visually and mentally the score;
2) Supervise every group of instruments (percussions, brass, wind, strings) both as individuals and as groups, paying attention if the players are properly playing their single part as indicated and rehearsed previously, according to the director's conception about the work played;
3) Supervise himself while conducting; is my right hand indicating the right tempo (velocity)? Is my left hand indicating the right dynamics (sound volume) to the various groups of instruments in different moments of the music? Is my facial and body posture suggesting the desired interpretation?;
4) Supervise the global effect of the actual execution, instantly comparing it with his personal ideas about the interpretation and acting properly to correct it when the actual playing is diverting from the initial concept;
5) In the same time a type of conscious splitting of the mind permits him to "feel" himself in the middle of the listeners, doing two actions contemporarily: valuating the musical result of the performance from the points of view of sonority (volume) and interpretation; and observing the reactions of the listeners to the whole act.

The conclusions of this "alter ego" (another me) are instantly analyzed by director's mind, determining his reactions as a whole, in a continuous cycle of actions-reactions. The professional slang call this phenomenon "to feel the listeners"; it is a phenomenon also well known to the actors, mainly in those with a direct contact with the audience (in cabaret) and to politicians addressing the people during a meeting.

This conscious mind- splitting phenomenon, a clear manifestation of the OBE, it is not simply only desired by the director; it is the result of the specific training of a musician (it comes after some decades of professional practice). The professionalism of the musician supposed and implies the out-of-body experience as an essential part of being artist. This experience is planned, enacted and controlled by the musician's mind; an implicit example of the existence of the consciousness (cogito, ergo sum- I am thinking, therefore I am).(3)

2. The Inspiration

Frequently the inspiration is considered as a phenomenon hard to be defined and explained, originating somewhere, somehow, from some unknown superior entity. It is absolutely true; the inspiration comes from the most superior entity; its name is the human mind and it is the result of the brain's activity, mainly of the mechanics of memory.

Consider the fact that music is a language, like the English language, for example. A language is a phenomenon defined by four parameters: the pre-speech (visualization and vocalization); the vocabulary, made by all the words existing in a language; a grammar, which defines each word of the vocabulary (nouns, verbs, a.s.o.); and a syntax, which organizes the elements of the vocabulary as defined by the grammar, in logical structures (subject, predicate, sentence, phrase, period, a.s.o.).(4)

Pre-speech

In Music are present both aspects of the pre-speech: visualization and vocalization.

Visualization is represented by two elements: the physical expressivity and the written musical language. The physical expressivity is given by the movements of the body indicating the music parameters of: rhythm, meter, pitch, volume, speed. The written musical language implies not only the decoding of the information but also its mental representation (pitch, rhythm, meter, speed, volume; all simultaneously) and physical representation of the decoded information; this is essentially needed to direct and perform the music.

Vocalization: the interjections (hey, ho, wow, a.s.o.) are primeval direct expressions of the thought. They are common to the spoken language and music. Musical language combines these interjections using different pitches and organizing them according to rhythm cells and various meters. As in the spoken language whole propositions can be created, with an accomplished sense. Example: Hey! Ho! Mmmmmm! Woooow! Hoa! (in different pitches and rhythms) with the meaning of: You! Stop! Let me see! Wonderful!!.Go!

A classical example in music is the choral suite "The Cries of London" for mixed choir and instruments (sixteenth-seventeenth century) by William Cobbold, Richard Dering, Michael East, Orlando Gibbons and Thomas Ravenscroft. (also treated by Luciano Berio and Ralph Vaughan Williams in the twentieth century).

The pre-speech is the most important testimony of the common origin of music and spoken language; the interjections (vocalizations) are common to all mankind, with quite the same significance in all the languages.

If you are driving, wherever in the world, and see in the middle of the road a person with a hand raised and saying "Ho"! you will brake. (mainly if in the other hand the person holds a gun, but this is another story...)

On the cortex, the centers responsible for speech and audio information are not only closed in space, but also directly related. In the brain, research demonstrates that the brain's inferior pair of corpora quadrigemina forms a selective audio system, with faster and more complex reactions than the visual system.

The inferior corpora quadrigemina are connected with the reticular formation, the Cajal's interstitial nucleus and the central gray matter. The system acts directly on the CNS and ANS and, following their ramifications, on the whole body.

The audio information is received, analyzed and transmitted to the different body's systems faster than the visual information. In turn, the body's feedback at the audio information is very fast and elicits a continuous process (feed forward/feedback).

Think for example at the reaction of the listeners at a dance or a ritual music rhythm: the body starts moving following the rhythm after a very short period of time, as a demonstration of the validity of the above statement.

Vocabulary: as any language, Music has its dialects. The sounds forming the vocabulary could be: natural (from the environment), mechanically produced (by musical instruments except the MIDI instruments), vocal, and electronically produced sounds (by sound generators, MIDI systems, PC produced sounds).These sounds are classified by the grammar and organized by syntax according to their specificity.

Grammar: in Music, the Grammar organizes (like in spoken language) the elements of the vocabulary, defining them as part of proposition (nouns, verbs, a.s.o.)

Syntax: is the way in which the propositions are organized in order to form a logical development of the content (subject, predicate, a.s.o.). The parameters of Spatial and Temporal dimensions of the organization of sounds build Musical Form; it is the music as is written, played and perceived.

Musical Form constructed in space and developing in time is usually both a result and a goal. It is the result of a compositional process having as its goal to express the thoughts and feelings of the author and to obtain the reaction of the listeners. Like in other products of human ingeniousness there is a direct rapport with the composer's personality, professional skills and artistic goals.

Musical form represents to music what the rules of grammar and syntax represents for the language: a logical organization of its components. The perception of music itself is multidimensional; the musical form creates in the mind a clear image of a logical expression. Observing for example the simple three–parts musical form ABA, if consider the links between music and consciousness we should take in consideration not only the spatial and temporal development of the musical material but also the image it creates in the mind of the listener by means of memory's mechanics.

An important contribution to understanding the mechanics of memory appears in the last thirty years with the introduction of the inter-disciplinary concept of meme. Dawkins (1976) coined the term as an analogy to the biological unit of inheritance, making a distinction between genetic replicants (genes) and non- genetic replicants (memes).(5) The meme is the smallest information pattern, that replicates affecting human mind and causing the propagation of the pattern.

Hofstadter (1985) considers that all transmitted knowledge is memetic. Another definition of meme, closer to the significance this concept has in the study of consciousness, was given by Wilkins (1998). He consider the meme as "the least unit of socio- cultural information relative to a selection process that has favorable or unfavorable selection bias that exceeds its endogenous tendency to change". The existence of musical meme (Vaneechoutte 1998,) is of a basically importance in understanding the consciousness and in composing music for therapy. Rhythm- memes alone or associate to sound-pitched memes are common to all mankind. The differences regarding cultural or social environments and heritage are reducible to basic elements, common to all cultures (this is one of the particularities of musical memes which difference them from another types of memes). To notice that, at least for the history of musical research, this concept appears initially in the late 1920 in Eastern Europe; it was studied and applied with very useful results by Bartok (1927)(6) and Brailoiu (1930)(7). Elements of musical folklore are common to all mankind – especially the rhythmical and melodic patterns related to the great events in life like birth, matrimony, death. Children's folklore presents the same basic patterns (memes) in Europe, Asia, America, Australia, Africa.(8) Beside this aspect there are physiological, anthropological and psychological details to sustain this theory. Meme is mainly universal and intrinsically to human mind, sub-memes strictly related to social groups or specific cultures are in a much lesser number. Memes are a specific result of the cultural development of the human beings' mind along the time; they created and developed not only the single individual's personality, but also the culture and behavior of the whole human society with the social relationships implied. That's why only the humans could perceive and recognized Music as such; the lab animals perceiving the same musical information react to it as to environmental sound/s, reacting only to the patterns of rhythm and pitches. This is the way the circus animals are trained, from the dawn of mankind, by associating the rhythm patterns to determined actions. The melody, harmony, orchestration, the musical structure itself, what defines the Music as a physical, psychological and cultural phenomenon has no significance for the animals; missing these elements the consciousness as human mind's aspect and the virtual therapeutic effects of Music are inexistent.

This is the reason why much laboratory research on "music and..." (neuroscience, brain, mind, a.s.o.) describes only the physiological mechanics of the perception of sounds in lab animals' brain, not of Music and its effects on the human brain. Music is Music only for the human beings, due to the mechanics of memory and cultural development of the mankind. No others living beings could experiment the Music's effects. No valid and useful data for understanding the reactions of a human brain to the Music could be obtained by studying rats- and no useful data for further therapy with human beings. Human beings just simple have MINDS and Consciousness.

THE KNOWLEDGE OF THE MUSICAL LANGUAGE

There are different levels of the knowledge of a language. The basic four levels are: the functional, the average, the conceptual and the creative levels.(4) The differences between them consist in an enlarged vocabulary and a better knowledge and use of the rules of grammar and syntax.

Example of functional level: suppose you meet in the street an alien. With the right hand raised (which is pre-speech, visualization) he (it) stopped you and said: "Mee.....hungry....where to....eat?" So, you explain to him/it how to reach the first burger shop (they accept whatever currency in the universe). The alien thanks you and gone away. What happened? The alien was able to express his need, understood the answer and acted accordingly. This is an example of the functional level of a language: the ability to express the feelings ("hungry"), the needs ("where to eat?") shortly, to communicate with the others with few words, without necessarily following the rules of grammar or syntax.

In Music, the functional level is represented by persons with the basic skills in playing an instrument (better a MIDI keyboard, a guitar or drums) or singing a song; or, more often than not, possessing a huge collection of musical recordings. Which made them believe they are expert musicians (they can tell to anybody who the composer is, and in which year the recording was made and by whom and where).Never present these persons with a real music score: in 99% of cases they are not able to read it; the answer usually is: "well, I need to hear it". Something likes a professional literature reviewer who needs to listen to the audio book version of a novel, for to have an opinion about the work. Only, they considered themselves "musicians".

The music composer translates into musical language his experiences (seeing a landscape, participating in an action, the ideas and images of a poem or a novel) both as they are perceived or understand by him and as the reactions of the composer to them. Exactly as a painter translate all in the visual language of the art; or a writer does in literature. Better and deeper the knowledge a composer has of the music compositional techniques and the science of music, richer will be his translation of the reality in the musical language.

The inspiration is a concrete, tangible phenomenon; it starts from the reality, even from a literary one. But from the moment it was translated into music (written and performed) and it is listening to by the listeners, it became abstract. In the example we had before, the Maestro starts from the image of the Moon in writing his Sonata; but in the moment the score is finished and the Sonata played, the music became abstract; the same music could suggest to the listeners many different images, often very different between them.

In order to experience the real phenomenon, try to do this experiment: listen to the recording of the music of a movie, without looking for the title of it. Even if you don't know the title, the plot of the movie, or the specific sequence for which the music was written, you can appreciate and enjoy the music of the sound track. You can also imagine situations very different of what the actual movie shown or without any ties with the movie. This process is a metasymbolic inspirational continuum, virtually infinite (the music creates in the mind an image which creates an image, which creates an image, and so on). Try to make this experience listening to the sound track of a foreign movie, in a language that you don't know; you'll appreciate the music even completely ignoring all about the movie, starting with the title.

In this process the mechanics of memory are determinant; the personal cultural background and the specific personality of the listener are the result of the functioning of various stages and processes of the memory.

3. The Improvisation

In studying the music a student (regular or self-taught) follows these steps:

1) Learn how the instrument is build and how it works;
2) Learn first of all the single parts (voices) of the score. For example in a piano piece, using both hands, the first to be studied is the right hand's part, then the left hand's part; finally both parts are played together.

What happened in this case? The student learns and memorizes patterns of complex relationships. The relationships between the written representation of the sound (the score), the real sound (the played note), the specific tool used to obtain the sound (the fingering in the piano's case), the physiological mechanics necessary to produce the sound (in the case of piano the various positions of the arm and different pressure to be applied by each muscle of it and the fingers); the logical order of the execution (the fingers number 1,3,5 of the right hand in playing a C major chord at piano, for example).

These patterns are stored in the long term memory banks. Every time when a similar situation is presented the hand of the player will take the memorized position of the specific pattern; the continuous practice of an instrument has exactly this goal, to create conditioned reflexes context- related. In time, these patterns become tools necessary to express the conceptions of the artist, are the basic of his interpretative art and determine his personal creativity.

When a musician "improvise" starting from a given theme or "freely", his mind consults the long term memory data banks and determines the reaction of the organism to every mental stimulus both in detail (what configuration the hand of the piano player will take during the performance) and regarding the general architecture of the music (the musical thinking in its space-time evolution).

Consider for example a jazz performer. He will first memorize the technical aspects peculiar to a determined style (how to play with the left hand a bass figure in ragtime or swing style, for example); every time when he will have to develop a theme in the specific style his hand will play the pattern of complex relationships described above.

Actually, such a thing like absolute improvisation cannot exist. Ex nihilo, nihil (nothing come from nothing). What we call with certain ignorance "improvisation" it is nothing more than the capacity of the player to combine in a new way patterns of relationships previously memorized. Sometime with very interesting results from the point of view of the expression and content; but one can always recognize the various patterns present in the creation of the music.

Take the case of the self-taught musicians, as usually the rockers and alike are; they are learning these patterns by imitating their favorite artists, often ignoring the musical written language.

Their creativity is absolutely necessary limited to the possible logical combinations between the basic patterns. Reason why, more often than not, the listeners comment: "I know this music from somewhere. I heard it before... it sounds like...it is copied from..." Not necessary the music was copied; but the fact that it is the result of a new combination of pre-existent patterns (as working with Lego modules) gives to the music a feeling of "heard before" (the mechanics of memory are the same for all, listeners enclosed).

In improvisation's case the relationships between music and consciousness are fundamental; the consciousness reveals itself in one of its basic aspects: it is the result of mechanics of memory in their full function.

Let consider now some theoretical aspects, with social and cultural implications, of the relationships between music and consciousness.

MUSIC AND CONSCIOUSNESS: THEORETICAL ASPECTS

1.Qualia

The concept of qualia appears first in C.S.Lewis' work "Mind and the World Order" published in 1929. Etymologically, the meaning is "qualia"-plural of "quale", in Latin; "of what sort" in English. It is defined as a quality regarded as an independent object and a feeling having a distinct quality.(9)

An old Sufi story which describes very well the concept of qualia: a person asked another "What is like, coffee? What did you experience when you drink it?" The answer: "If you want to know what is like coffee, just drink it!".

A similar question regarding Music could be "What did you experience when you're listening to the music of Mozart? (or Metalica, it is just the same in this case).

The description of direct and personal reactions when experiencing a phenomenon and the definition of these reactions as self-explanatory phenomena it is the matter of the concept of qualia. The concept itself is a demonstration of the existence of consciousness: I am asking myself, therefore I am thinking (dubito, ergo cogito); I am thinking, therefore I am (cogito, ergo sum).

Do musical qualia exist? The answer is yes and not.

Music is a complex phenomenon, as mentioned before; and as any phenomenon it has causes and effects.

Let's consider two general aspects of the Music; the producers and the consumers.

The producers, composer and interprets, are the very source of the music phenomenon; they know perfectly what Music is, how it is produced and what probably effects the Music has. Evidently, for them the concept of qualia doesn't make sense: they understand perfectly what Mozart's or Metalica's music is, in all its aspects; they are making a living with this.

Even when the producers became consumers, by listening to a recording or assisting to a concert, the things doesn't change; professionals are listening in a different way, the mind analyzing both the music architecture and music interpretation from the point of view of the science of music; perfectly understanding the way music is build and performed, predicting the possible effects of it.

Considering **the causes** of the Music phenomenon the musical qualia doesn't and cannot exist at all.

Another matter is if we consider the consumers, not professional musicians; which means the majority of the listeners. In their case we have to deal with **the effects** of the phenomenon, not with its causes.

From this point of view the musical qualia exists; their type and number is as big as the number of the listeners, an example of the Latin locution "Quod capita, tot sensus" (there are as many ideas as persons are).

The same music elicits different reactions in different listeners. The same cause produces different effects. The mechanic of Brain-Mind-Brain- Organism system, mentioned before, is acting in different ways in the consumers than in producers. The mechanics of memory in case of consumers are using exclusively the long term memory data banks, the so-called "individual culture".

This "individual culture" consists in a quantity of data (notions) organized in direct and crossed patterns of relationships. The mind analyzes the musical information comparing it with the existing data stored in the memory, trying to find out differences and similitude of various types. The conclusion could be, for example: "This music has a structure similar to one stored in my memory; it belongs to that type and style of music, probably to that specific composer; partial conclusion: I like it". Final conclusion: relax, is Mozart (or Metalica, accordingly).

The musical qualia exist only if related to **the effects** of music listening upon the consumers, not professional musicians; which are the majority of people, with their individual and collective consciousness, stimulated and activated by the effects of music listening.

2. Music Aesthetics

Music aesthetics are dealing quite exclusively with the effects of music listening; musical qualia are described well and by large by various authors, starting with the first decades of the nineteenth century; the consumers need the comments of other consumers about the product they like to enjoy. Few producers wrote about musical qualia, and only in order to stress their opinions about music composition, giving always the explanations for their technical decisions in order to let the consumers understand better how the product is made. Among this producer's category there are personalities like Berlioz, Messiaen, Honegger, Strawinsky; very important authors in the music's history and a clear example of the differences about how Music is understand by the two categories of producers and consumers.

Apparently there is a huge dichotomy between the two categories; but the gap is only apparent. Considering only the effects of the phenomenon could lead to partial conclusions; often these partial conclusions became viruses of the mind and conditioned the general opinions about Music and its role in the society's structure. Concepts as "art for art, pure art" versus "art with tendency, art with social goals" are both related to the effects of the Music. It could seem that these concepts are abstract and not important for the large mass of the listeners; history shown that it could lead to the oppression of the producers and in some cases even to their physical elimination. All the twentieth century's dictatorships directed their social and cultural politics toward the implementation of the concept of "art with tendency" in order to oblige the musicians to serve their ideology. Composers like Shostakowitsch, Prokofiev, Kurt Weil and many others were victims of this reality.

The fascist Italy, Hitlerian Germany, communist USSR and the communist-occupied countries in the Eastern Europe, the Maoist China, are places in the world where aesthetics became official policy and lead to moral and physical crimes. Ignoring the causes of the phenomenon leads to un-realistic conclusions; but these conclusions doesn't remain in the realm of abstract concepts, became very realistic policies with thousands of victims; and viruses of the collective mind.

The fact that the same music (which is abstract) could be used for a symphony or for an ideologically "correct" song by just adding the appropriate lyrics, from a part shown the ignorance by the rulers of what Music is; from the other part helped the producers to survive, sometimes by just giving an "ideologically correct" title to their music ("Long Live the Leader Symphony", for example).(10)

The Music phenomenon could be considered from many points of view; but it must be take in consideration the reality of causes and effects. Ignoring the first could lead to a "sky is the limit" un-realistic effect.

The relationships between two complex phenomena like Music and Consciousness are as complex as the phenomena are. In this paper I tried to analyze some of the most important among them, hoping that the readers will be interested to discover the deepest reality beyond the apparent sentence "Music is an Art", period.

REFERENCES

1. Graur, Alexander : *Music Integrative Neurotherapy™, Part One: Body and Sound;* 2003, MPD, Medicamus Publishing Division, USA/Italy

2. Blackmore, S. (1982**). Beyond the body: an investigation of out-of-the-body_experiences**. Granada, London.

3. Cartesio, **Discorso sul metodo**. Laterza, Roma-Bari, 2004.

4. Chomsky, Noam (1959). "Reviews: _Verbal behavior_ by B. F. Skinner". _Language_. **35** (1): 26–58. JSTOR 411334.

5. Dawkins, Richard *(1989),* "11. *Memes: the new replicators*", The Selfish Gene *(2ⁿᵈ ed., new ed.),* Oxford: Oxford University Press, *p. 368,* ISBN 0-19-217773-7

6. *Kroo, Gyorgy. Guide to Bartok. Branden Publishing Co. ISBN 978-0-8283-1559-3*.

7. *Brailoiu, Constantin:Folklore musical.* Encyclopédie de la musique Fasquelle, Paris 1959.

8. Iusceanu, Victor: Moduri si game, 1962, Editura Muzicala Bucuresti, Romania

9. Simon Blackburn in _The Oxford Dictionary of Philosophy_ Second edition

10. Volkov, Solomon: *Shostakovich and Stalin: The Extraordinary Relationship Between the Great Composer and the Brutal Dictator.* Knopf 2004.

In the context of this article trance is meant to be a state of consciousness that is devoid of all reality although it seems real. In my research I realized that the impression of reality is an illusion induced by a perfect interaction of the sensory perception and this illusion is coming, going and changing its appearance. Only by identifying with it, as soon as we sense ourselves as separate identities, we belief that to be real. It is disappearing as soon as awareness is focussed on and therefore I call it a trance.

CAN CONSCIOUSNESS INFLUENCE OUR EPIGENETICS AND CAN EPIGENETICS INFLUENCE OUR CONSCIOUSNESS?

Ingrid Fredriksson

ABSTRACT EPIGENETICS IS A mechanism for regulating gene activity independent of DNA sequence that determines which genes are turned on or off: in a particular cell type, in a different disease states or in response to a physiological or even psychological stimulus.

There is a microbiota–gut–brain axis communication in health and disease. (Under healthy conditions, the predominance of symbiotic bacteria, an intact intestinal barrier, a healthy innate immunity controlling pathobiont overgrowth inside the intestinal barrier).

The molecules that constitute epigenomes have no resemblance of DNA. While DNA is a double spiral, similar to a twisted rope ladder, the epigenome is a system of chemical markers that sits on the DNA. What is its purpose? In the same manner that a conductor leads an orchestra, the epigenome decides how the genetic information of DNA shall be should be expressed. The molecule markers either engage or disengage the genes depending upon the cell's needs and environmental factors, such as diet, stress and poisons. Of late, the discoveries surrounding the epigenome have caused a revolution in the field of biology now being able to prove a connection between the epigenome and certain illnesses, including aging.

Keywords Consciousness, Epigenetics, DNA, cells, genes, microbiota, gut and brain

INTRODUCTION

In our body there are 100 trillion (10^{12}) cells each of which containing circa 22 000 genes. Negative epigenetic changes can increase the risk of illness while positive epigenetic changes minimise the risk for illness. From different methyl–donators, such as methionine, folic acid, choline, betaine and vitamin B2, B6 and B12 SAM is formed as methylated DNA and histones. When the fiber breaks down in the large intestine with the help of probiotic bacteria, short chain fatty acid is the product, especially butyric acid that has the important task of acetylation of histones. A line of polyphenol/flavonoids/iridiods from berries, vegetables, green tea and dark chocolate with a high cocoa content may have strong epigenetic impact. Amongst these there are certain substances noticeable, genistein from soya, curcumin from turmeric, resveratrol from red wine and lingon berry, isothiocyanates from the cabbage family, allyl sulfides from onions, epigallocatechin gallate (EGCG) from green tea and a line of polyphenol/flavonoids/iridiods from lingon, blueberries and Hippophae rhamnoides to name but a few. All of these substances can methylate DNA and histones and acethylate histones. Normally it requires a great amount. Through the named substances it is also possible to get the positive effects of non-coded RNA (ncRNA; miRNA). Physical activity, including working out, and how we feel, can also give you positive epigenetic benefits.

New mechanisms found in epigenetics open new possibilities for consciousness, whatever it is, to interact with the physical reality.

EPIGENETICS

Epigenetics is a mechanism for regulating gene activity independent of DNA sequence that determines which genes are turned on or off:

in a particular type
in different disease states
in response to physiological or psychological stimulus.

All of our cells have the same DNA-sequence but they look different and function differently. In all cells there are similar programs for mutual functions e.g. cell division, cell movements, cell death and forming of energy (ATP). Through epigenetic mechanisms, different DNA-sequences are started or stopped which is also the reason for their special functions.

The connecting or disconnecting of methyl groups to or from DNA and histones is the most studied mechanism in epigenetics. Methylation of disables transcriptions factors which cannot bind themselves to DNA which in turn stops transcription.

More recently researchers become aware of another set of mechanisms in cells – epigenome, a word that can be translated to "higher genome". Epigenetics is the study of this wondrous set of chemical mechanisms.

The molecules that constitute epigenomes have no resemblance of DNA. While DNA is a double spiral, similar to a twisted rope ladder, the epigenome is a system of chemical markers that sits on the DNA. What is its purpose? In the same manner that a conductor leads an orchestra, the epigenome decides how the genetic information of DNA shall be should be expressed. The molecule markers either engage or disengage the genes depending upon the cell's needs and environmental factors, such as diet, stress and poisons. Of late, the discoveries surrounding the epigenome have caused a revolution in the field of biology now being able to prove a connection between the epigenome and certain illnesses, including aging.[1]

The last piece of a chromosome is called a telomere. Its length degreases a little with each cell division until the telomere reaches a certain length at which division is no longer possible, whereupon the cell ages and dies. This is a normal course of events. A cancer cell, however, triggers a gene that instructs the cell to create an enzyme, telomerase, which prevents the telomeres from getting too short in the course of cell division. The discovery of telomerase provides new possibilities for stopping cancer growth. If the gene that shuts off the production of telomerase is found, cancer cells could be prevented from continuing to divide. It was feared at first that ordinary cells would be converted into cancer cells if they were exposed to telomerase. But that is not the case. [2]

For their discovery of how chromosomes are protected by telomeres and the enzyme telomerase, Elizabeth Blackburn, Carol Greider and Jack Szostak received, 2009, the Nobel prize in Physiology or Medicine. Blackburn has, as opposed to many other receivers of the Nobel prize continued with research, presently in the quite spectacular subject of meditation. She has concluded that meditation lengthens the telomeres leading to longer life through lowering blood pressure, reducing infections and dangerous stress. Lab studies show that the stress hormone cortisol reduces the activity of telomerase, while oxidative stress and inflammation – the physiological fallout of psychological stress – appear to erode telomeres directly. Age-related conditions from osteoarthritis, diabetes and obesity to heart disease, Alzheimer's and stroke have all been linked to short telomeres. [3]

Exercise boosts telomere transcription. Endurance exercise and metabolism are linked to transcriptional activation of human telomeres, researches propose. "There is a new link between metabolism and telomeres," said study coauthor Anabelle Decottignies of the Duvc Institute in Brussels, Belgium. [4] When healthy individuals perform a cardiovascular workout, their muscles increase transcription of telomeres, according to a study published in *Science Advances*. The team also identifies a novel transcription factor that appears to promote telomere transcription and provides the first direct evidence that telomere transcription is linked to exercise and metabolism in people.

The majority of normal cells do not divide themselves when DNA becomes damaged. [5] Their chromosomes therefore run the risk of becoming too short and they have no need for a high level of telomere activity. Cancer cells, on the other hand, have the ability to divide themselves indefinitely. How can they then preserve their telomeres and avoid cell aging? It has been discovered that many cancer cells have a high level of telomere activity, and that this has increased the hope that it be possible to treat cancer through removing the telomerase. At present studies of the telomerase are being carried out as well as clinical experiments which test vaccines that are aimed at cells with an increased telomerase activity.

Certain hereditary illnesses have been shown to depend upon telomerase defects, one of these is certain types of hereditary aplastic anemia where insufficient cell division activity in the bone marrow's stem cells causes a severe lack of blood.

Other examples are certain hereditary skin and lung diseases.

To sum up, these discoveries have not only given us a new picture of the cells mechanisms but have also brought attention to important medical problems thus stimulating the development of new possible treatment [6]

Chromosomes are constructed of chromatin which is a composition of DNA and proteins, mainly histones but also other gene regulating proteins. The DNA molecule is made up of two long chains of simple cells, aka nucleotides. Nucleosomes are the smallest repeating units in the chromatin to construct chromosomes. Nucleosomes are made up of DNA with circa 145-147 base pairs in length coiled around the nucleosome nucleus which consists of 4 pairs of rounded histone proteins. Through a electron microscope the nucleosomes look similar to "pearls on a string of DNA". The histones H2A, H2B, H3 and H4, which there are two of in each histone, make up the nucleus of the nucleosome. [7] Surrounding this is the DNA while the histone H1 is situated in the position where histone bound DNA goes on to the "DNA-link" i.e. DNA which links one nucleosome with another. The histones N-terminal tails stick out from the nucleosome structure and are therefore chemically accessible to the surroundings.

The known epigenetic mechanisms are 1. DNA-methylation 2. Histone modifying. 3. Chromatin regulating proteins. 4. Chromatin structure (chromatin landscape). 5. Imprinting. 6. RNA-mediated mechanisms (not long coded RNA. Micro-RNA;miR). The following concerns particular DNA-methylation and histone modification. Imprinting means inactivation of either a certain gen from the mother or the father. [8]

DNA-METHYLATION

DNA-methylation is the most studied epigenetic mechanism. Methylation leads to the silencing of the gen. DNA-methylation occurs particular of cytosines followed by guanosine, aka CpG-nucleotides, with the construction of 5-methyl-cytosines. DNA-methylation is controlled by three different DNA-methyltransferase (DNMTs) coded by different genes on different chromosomes: DNMT1, DNMT3a and DNMT3b.[9] Examples of such can give DNA-methylations are food that we eat, medicines we take, poisons we are exposed to or taxing of the muscles. Of importance in this context is, methylations-mechanisms with folic-acid, vitamin B6, Vitamin B1 and choline. Increased homocysteine can interrupt these functions.

Negative epigenetic changes can increase the risk of illness while positive epigenetic changes minimise the risk for illness. From different methyl-donators, such as methionine, folic acid, choline, betaine and vitamin B2, B6 and B12 SAM is formed as methylated DNA and histones. When the fiber breaks down in the large intestine with the help of probiotic bacteria, short chain fatty acid is the product, especially butyric acid that has the important task of acetylation of histones. A line of polyphenol/flavonoids/iridiods from berries, vegetables, green tea and dark chocolate with a high cocoa content may have strong epigenetic impact. Amongst these there are certain substances noticeable, genistein from soya, curcumin from turmeric, resveratrol from red wine and lingon berry, isothiocyanates from the cabbage family, allyl sulfides from onions, epigallocatechin gallate (EGCG) from green tea and a line of polyphenol/flavonoids/iridiods from lingon, blueberries and Hippophaer hamnoides to name but a few. All of these substances can methylate DNA and histones and acethylate histones. Normally it requires a great amount. Through the named substances it is also possible to get the positive effects of non-coded RNA (ncRNA; miRNA). Physical activity, including working out, can also give you positive epigenetic benefits.

Epigenetics, greatly simplified, is concerned with three main mechanisms:

1) The attachment or releasing of methyl-groups (CH3-) onto the DNA or on the nine histones (methylation respectively demethylation). Methylation of DNA means the quieting of the genes. The histones are the "bullet" that the double lined DNA-molecule is circulated around. Histones N-terminal tail stick out and are therefore chemically available to the surrounding. Methylation of lysine relates to the most usual. For example, Trimethylation of lysine 27 or lysine 9 on histone H3 connected to the silencing of the genes.

2) The attachment or the removal of the acetyl group (HC3-COO-) plus other substances (acetylation and deacetylation) of the histones.

3) The effect of non-coding RNA; ncRNA, especially microRNA (miRNA) which can stop the building of a specific protein. [10]

Choline – A vital nutrient

Choline was for acknowledged officially in 1998 as a vital nutrient necessary for normal cell function.

Cholin is needed for:

1) Synthesis of the signal substance acethylcholine (Ach)
2) The cells membranes structure (phospholipid, sphyingomyelin)
3) Cell membrane signaling (synthesis of substances from phospholipid)
4) Mitochondrion function
5) Methyl groups circulation (reduction of Homocysteine to methionine thereafter the forming of S-adenocylmethionine; SAM, that is that ultimate donator to the methyl group.

In large parts of the Western world the recommended intake of choline is not adhered to. A study showed that only 10 % of the USA's population received enough choline. Other studies show instead 20-25 %. A lack of choline can cause increased levels of homocystein bearing with it negative consequences (mainly heart and blood vessel illnesses). A shortage of choline may also lead to a rupture of the DNA-string and changes epigenetic markers on the DNA and histones.

Where to find choline? The richest sources of choline are of animal origin, especially egg yolk and liver. Other sources are meat, milk, wheat germs, soya nuts and legumes. There is a number of food supplements with choline. Many of these also contain a certain amount of B vitamins B2, B6 and folate.

Inflammation conductors are for example: Milk powder, sugar, simple carbohydrates, margarine and biscuits. The opposite, healthy food:

Fibers (ones that form butyric acid, butyrate together with probiotic bacteria in the large intestine), Species of the cabbage family, Nitrate enriched plants such as beetroot, Folic acid B6, B12, B2 – methionine, Choline (Egg yolk, liver, meat, milk, wheat germ, soya, nuts, legumes), Glutamine, Vitamin D, Fatty fish – omega 3, and selen for example. [1]

- Gut microbiota plays an important role in our lives and in the way our bodies function.
- The composition of gut microbiota is unique to each individual, just like our fingerprints.
- Our gut microbiota contains tens of trillions of bacteria – ten times more cells than in our body.
- There are more than 3 millions microbial genes in our gut microbiota –150 times more genes than in the human genome.
- Microbiota, in total, can weigh up to 2 kg.More than 1,000 different known bacterial species can be found in human gut

microbiota, but only 150 to 170 predominate in any given subject.

The word microbiota represents an ensemble of microorganisms that resides in a previously established environment. Human beings have clusters of bacteria in different parts of the body, such as in the surface or deep layers of skin (skin microbiota), the mouth (oral microbiota), the vagina (vaginal microbiota), and so on.

Some of the functions are:

It helps the body to digest certain foods that the stomach and small intestine have not been able to digest.

It helps with the production of some vitamins (B and K).

It helps us combat aggressions from other microorganisms, maintaining the wholeness of the intestinal mucosa.

It plays an important role in the immune system, performing a barrier effect.

A healthy and balanced gut microbiota is key to ensuring proper digestive functioning.

Taking into account the major role gut microbiota plays in the normal functioning of the body and the different functions it accomplishes, experts nowadays consider it as an "organ". However, it is an "acquired" organ, as babies are born sterile; that is, intestine colonisation starts right after birth and evolves as we grow.

Sterile inside the uterus, the newborn's digestive tract is quickly colonised by microorganisms from the mother (vaginal, faecal, skin, breast, etc.), the environment in which the delivery takes place, the air, etc. From the third day, the composition of the intestinal flora is directly dependent on how the infant is fed: breastfed babies' gut microbiota, for example, is mainly dominated by Bifidobacteria, compared to babies nourished with infant formulas. Scientists consider that by the age of 3, microbiota becomes stable and similar to that of adults, continuing its evolution at a stages rate throughout life.

Gut microbiota's balance can be affected during the ageing process and, consequently, the elderly have substantially different microbiota to younger adults.

While the general composition of the intestinal microbiota is similar in most healthy people, the species composition is highly personalized and largely determined by our environment and our diet. The composition of gut microbiota may become accustomed to dietary components, either temporarily or permanently. Japanese people, for example, can digest seaweeds (part of their daily diet) thanks to specific enzymes that their microbiota has acquired from marine bacteria.

Although it can adapt to change, a loss of balance in gut microbiota may arise in some specific situations. This is called dysbiosis. Dysbiosis may be linked to health problems such as functional bowel disorders, inflammatory bowel disease, allergies, obesity and diabetes.

Many studies have demonstrated the beneficial effects of prebiotics and probiotics on our gut microbiota. Serving as "food" for beneficial bacteria, prebiotics help improve the functioning of microbiota while allowing the growth and activity of some "good" bacteria. Present in some fermented products such as yoghourt, probiotics help gut microbiota keep its balance, integrity and diversity.

Thanks to technological progress, the picture of the bacteria living in the gastrointestinal tract is becoming clearer. Researchers now use a range of techniques, including the tools derived from molecular biology, to further clarify the mysteries of microbiota. While there are still some things that are yet to be discovered, more and more findings are being presented every day.

New findings indicate that bacteria in the intestines not only degrade food but also can communicate with parts of the brain and thereby affect the appetite and burning off of fat. Perhaps this occurs via the hormones, signaling substances in the blood, or via nerves that go from the intestines to the brain. According to studies carried out on mice it seems as though bacteria in the intestines can affect behavior e.g. how much and for how long the mouse eats. Now, researchers at Sahlgrenska hospital investigate what types of differences between individual mice gut flora which can affect the brain and therefore appetite and fat burning. With time, the purpose of the research will be to ascertain as to whether the results are applicable to humans, and if a person's gut flora will give them a "hungry brain" which may increase the risk for obesity. "This is another example of how important intestinal bacteria are for different biological functions" says professor Fredrik Bäckhed.

The gut microbiota reduces leptin sensitivity and the expression of the obesity suppressing neuropeptides proglucagon (Gcg) and brain-derived neurotrophic factor (Bdnf) in the central nervous system to be published in Endocrinology.[11] This might be the way, bacteria communicate with each other!

"The never ending conflict of "nature versus nurture" loses some of its relevance when nature and nurture can no longer be differentiated. Where is the line drawn for the individual's moral responsibility? Is it at the womb, or perhaps at the first, second or third trimester or generation? Through mapping the mechanisms of how our way of life and our environment, which essentially is stipulated by our current economic situation, causes sickness and human suffering, how should we then approach these constitutions? Is it the role of medical science and doctor's to passively allow disease to develop only to then concentrate on controlling it by means of medicinal remedies? Remedies that despite, the name seldom heal." [12]

The World Health Organization estimated that more than 600 million people in the world are obese. Although studies have revealed genetic contributors to weight, the hereditary component of obesity is not well understood. The results of a study published (January 28) in Cell have suggested that a person's predisposition to obesity is at least partially determined by epigenetic regulation. [13]

"We show that you can have a strong phenotype [obesity] with absolutely no genetic underpinnings," study coauthor J. Andrew Pospisilik of the Max Planck Institute of Immunobiology and Epigenetics in Freiburg, Germany, told Science.[14]

Previous work had shown that genetically identical mice with only one copy of *Trim28*—a gene that encodes a chromatin-interacting regulatory protein—showed considerable variation in weight. Generating their own populations of "identical" mutant mice, the researchers discovered that this variation was not normally distributed, but associated with two distinct groups: lean animals and obese ones.

When the team probed the mutants' gene-expression profiles, it found that mice in the obese group showed reduced expression of an interacting network of imprinted genes—those that are expressed by either the maternally or paternally inherited copy. The results suggest that the Trim28 protein forms part of an epigenetic switch, the authors wrote in their paper.

"Once the switch is triggered, it is a lifelong, epigenetically driven decision that ends in a stable, either a lean or obese phenotype," Pospisilik said in a press release. "The effect is akin to a light switch—on or off, lean or obese."

Turning to humans, the researchers found that *Trim28* expression was unusually low among obese children. What's more, publicly available datasets from genetically identical twins with different weight profiles suggested a similar relationship between gene expression and body mass index, supporting an epigenetic component to obesity.

Paul Franks of the Lund University Diabetes Center in Malmö, Sweden, who was not involved in the work, told *Science* the researchers now must identify the regulators of this network switch. "The big question now is, 'What is the trigger?'" he said. "If you could determine what that was, you'd have the basis for intervention."

About ten years ago, Sweatt's lab made the seminal discovery that everyday experiences tap into epigenetic mechanisms in sub regions of the brain, and the resulting epigenetic changes in DNA are critically important for long-term memory formation and the stable storage of long-term memory. The 2007 Neuron paper "Covalent modification of DNA regulates memory formation," by Courtney Miller, PhD, and Sweatt, was the first to show that active regulation of the chemical structure of DNA is involved in learning and experience-driven changes in the brain. [15] This work was supported by National Institutes of Health grants T32HL105349, MH57014, P60DK079626 and P30DK56336. The T32 pre-doctoral fellow grant to Heyward from the UAB Nutrition and Obesity Research Center supported his training in the biological basis of obesity.

THE DUTCH HONGERWINTER

During the winter of 1944-1945, Nazi Germany blockaded towns across Western Netherlands, a period which become known as the Dutch Hongerwinter. [16]

Many decades later, scientific research found that the children born during this famine were underweight and more likely to suffer from disease. What was most startling, however, was that these children's children were also born significantly underweight, despite never having experienced a nutritional scarcity during in vitro development. Researchers had concluded that the famine "scarred" the DNA of the victims, but it was only recently that we were able to correctly identify this "scarring" as epigenetics.

Other studies have proposed a more tentative connection between one generation's experience and the next. For example, girls born to Dutch women who were pregnant during a severe famine at the end of the second world war had an above-average risk of developing schizophrenia. Likewise, another study has showed that men who smoked before puberty fathered heavier sons than those who smoked after. The team were specifically interested in one region of a gene associated with the regulation of stress hormones, which is known to be affected by trauma. "It makes sense to look at this gene."

They found epigenetic tags on the very same part of this gene in both the Holocaust survivors and their offspring, the same correlation was not found in any of the control group and their children. Through further genetic analysis, the team ruled out the possibility that the epigenetic changes were a result of trauma that the children had experienced themselves.

"To our knowledge, this provides the first demonstration of transmission of pre-conception stress effects resulting in epigenetic changes in both the exposed parents and their offspring in humans," said Yehuda, whose work was published in Biological Psychiatry.

It's still not clear how these tags might be passed from parent to child. Genetic information in sperm and eggs is not supposed to be affected by the environment – any epigenetic tags on DNA had been thought to be wiped clean soon after fertilization occurs.

However, research by Azim Surani at Cambridge University and colleagues, has recently shown that some epigenetic tags escape the cleaning process at fertilization, slipping through the net. It's not clear whether the gene changes found in the study would permanently affect the children's health, nor do the results upend any of our theories of evolution.

Whether the gene in question is switched on or off could have a tremendous impact on how much stress hormone is made and how we cope with stress, said Yehuda. "It's a lot to wrap our heads around. It's certainly an opportunity to learn a lot of important things about how we adapt to our environment and how we might pass on environmental resilience."

The impact of Holocaust survival on the next generation has been investigated for years – the challenge has been to show intergenerational effects are not just transmitted by social influences from the parents or regular genetic inheritance, said Marcus Pembrey, emeritus professor of pediatric genetics at University College London. [17]

"Yehuda's paper makes some useful progress. What we're getting here is the very beginnings of a understanding of how one generation responds to the experiences of the previous generation. It's fine-tuning the way your genes respond to the world."

Can you inherit a memory of trauma?

Researchers have already shown that certain fears might be inherited through generations, at least in animals.

Scientists at Emory University in Atlanta trained male mice to fear the smell of cherry blossom by pairing the smell with a small electric shock. Eventually the mice shuddered at the smell even when it was delivered on its own.

Despite never having encountered the smell of cherry blossom, the offspring of these mice had the same fearful response to the smell – shuddering when they came in contact with it. So too did some of their own offspring.

On the other hand, offspring of mice that had been conditioned to fear another smell, or mice who'd had no such conditioning had no fear of cherry blossom.

The fearful mice produced sperm which had fewer epigenetic tags on the gene responsible for producing receptors that sense cherry blossom. The pups themselves had an increased number of cherry blossom smell receptors in their brain, although how this led to them associating the smell with fear is still a mystery.

"The difference between these two is significant because this fundamental belief called genetic determinism literally means that our lives, which are defined as our physical, physiological and emotional behavioral traits, are controlled by the genetic code," Lipton said in an interview with the online magazine, *Super consciousness*. "This kind of belief system provides a visual picture of people being victims: If the genes control our life function, then our lives are being controlled by things outside of our ability to change them. This leads to victimization that the illnesses and diseases that run in families are propagated through the passing of genes associated with those attributes. Laboratory evidence shows this is not true." [18]

A STEADY DIET OF QUANTUM NUTRIENTS

"When we have negative emotions such as anger, anxiety and dislike or hate, or think negative thoughts such as 'I hate my job,' 'I don't like so and so' or 'Who does he think he is?', we experience stress and our energy reserves are redirected," an article on HMI's website explains.

This causes a portion of our energy reserves, which otherwise would be put to work maintaining, repairing and regenerating our complex biological systems, to instead confront the stresses these negative thoughts and feelings create.

"In contrast," the article continues, "when we activate the power of our hearts' commitment and intentionally have sincere feelings such as appreciation, care and love, we allow our hearts' electrical energy to work for us. Consciously choosing a core heart feeling over a negative one means instead of the drain and damage stress causes to our bodies' systems, we are renewed mentally, physically and emotionally. The more we do this the better we're able to ward off stress and energy drains in the future. Heartfelt positive feelings fortify our energy systems and nourish the body at the cellular level. At HeartMath we call these emotions quantum nutrients."

In simple terms most people can relate to, what this means is that when we are having a bad day, going through a rough period such as dealing with the sickness of a loved one or coping with financial troubles, we can actually influence our bodies – all the way down to the cellular level – by intentionally thinking positive thoughts and focusing on positive emotions.

CHANGING DNA THROUGH INTENTION

The power of intentional thoughts and emotions goes beyond theory at the HeartMath Institute. In a study, researchers have tested this idea and proven its veracity.

HeartMath researchers have gone so far as to show that physical aspects of DNA strands could be influenced by human intention. The article, Modulation of DNA Conformation by Heart-Focused Intention – McCraty, Atkinson, Tomasino, 2003 – describes experiments that achieved such results.

For example, an individual holding three DNA samples was directed to generate *heart coherence* – a beneficial state of mental, emotional and physical balance and harmony – with the aid of a

HeartMath technique that utilizes heart breathing and intentional positive emotions. The individual succeeded, as instructed, to intentionally and simultaneously unwind two of the DNA samples to different extents and leave the third unchanged.

"The results provide experimental evidence to support the hypothesis that aspects of the DNA molecule can be altered through intentionality," the article states. "The data indicate that when individuals are in a heart-focused, loving state and in a more coherent mode of physiological functioning, they have a greater ability to alter the conformation of DNA.

"Individuals capable of generating high ratios of heart coherence were able to alter DNA conformation according to their intention. … Control group participants showed low ratios of heart coherence and were unable to intentionally alter the conformation of DNA."

The influence or control individuals can have on their DNA – who and what they are and will become – is further illuminated in HeartMath founder Doc Childre's theory of *heart intelligence*. Childre postulates that "an energetic connection or coupling of information" occurs between the DNA in cells and higher dimensional structures – the higher self or spirit.

Childre further postulates, "The heart serves as a key access point through which information originating in the higher dimensional structures is coupled into the physical human system (including DNA), and that states of heart coherence generated through experiencing heartfelt positive emotions increase this coupling."

The heart, which generates a much stronger electromagnetic field than the brain's, provides the energetic field that binds together the higher dimensional structures and the body's many systems as well as its DNA.

Childre's theory of heart intelligence proposes that "individuals who are able to maintain states of heart coherence have increased coupling to the higher dimensional structures and would thus be more able to produce changes in the DNA."

MOM'S IN CONTROL—EVEN BEFORE YOU'RE BORN.

Researchers have uncovered previously unappreciated means by which epigenetic information contained in the egg influences the development of the placenta during pregnancy. The research, which was performed in mice, indicates that a mother's health, even before conception, may influence the health of her fetus, and opens questions on how a mother's age may influence placental development.

Epigenetic information is not encoded within the DNA sequence but is critical for determining which genes are on or off. One of the ways this is achieved is via DNA methylation, a biological process where the DNA is chemically tagged to silence genes. DNA methylation marks are laid down in each egg during their development in the ovaries and, after fertilization, some of these marks are passed onto the fetus and placenta.

In exploring the purpose of this maternal information in fetal development, focus so far has been on a small number of genes termed 'imprinted genes'. However, there are nearly one thousand other genomic regions where methylation in the egg cell is passed onto the early embryo. The researchers set out to explore the importance of this type of methylation on the development of the placenta, a vital organ in pregnancy, and their findings are presented in the latest issue of the journal *Developmental Cell*. [19]

"We were surprised to find that DNA methylation from the egg played a much larger role in placental development than methylation that was introduced after fertilization, whereas in the embryo both are important," explains Miguel Branco, a group leader from Queen Mary University of London who led the work. "Evolution, it seems, has granted mothers the tools to control the growth of their progeny during pregnancy by instructing on placental development."

By using mice in which methylation of the egg's DNA had been blocked, the researchers found that DNA methylation occurring during the development of the egg was essential for correct placental development. In particular, the research identified several genes regulated by methylation in the egg that are involved in cell adhesion and migration – both vital properties for cells of the developing placenta in establishing connections with maternal tissues to support embryo development.

"This was an exciting result for us," said Myriam Hemberger, a group leader at the Babraham Institute. "The phenomenon of gene imprinting explains some of this but our results show that the importance of DNA methylation in early development extends beyond imprinting. Specifically, maternally-inherited DNA methylation marks are important for normal placentation as they specify cellular properties such as adhesiveness and invasive character, as well as determining the correct balance of cell types needed in the placenta."

The research opens questions regarding the potential influence of maternal health on the fetus long before conception.

LICK YOUR RATS

Some mother rats spend a lot of time licking, grooming, and nursing their pups. Others seem to ignore their pups. Highly nurtured rat pups tend to grow up to be calm adults, while rat pups who receive little nurturing tend to grow up to be anxious.[20]

It turns out that the difference between a calm and an anxious rat is not genetic—it's epigenetic. The nurturing behavior of a mother rat during the first week of life shapes her pups' epigenomes. And the epigenetic pattern that mom establishes tends to stay put, even after the pups become adults.

In our society, we think of anxious behavior as being a disadvantage. But that's because, for the most part, we live in a nutrient-rich, low-danger environment. In the rat equivalent to our world, the relaxed rat lives a comfortable life. It is likely to reach a high social standing, and it doesn't have to worry about where its next meal is coming from. An anxious rat, on the other hand, doesn't do so well. It is more likely to have a low social standing and suffer from diabetes and heart disease.

In another environment, however, the tables turn. The anxious, guarded behavior of the low-nurtured rat is an advantage in an environment where food is scarce and danger is high. The low nurtured rat is more likely to keep a low profile and respond quickly to stress. In the same environment, a relaxed rat might be a little too relaxed. It may be more likely to let down its guard and be eaten by a predator.

We're used to thinking of inheritance in terms of the letters of the DNA code that pass to us from our parents — through eggs and sperm. But the rat-licking story tells us that there is another path to the offspring's DNA. Through her licking behavior, a mother rat can write information onto her pups' DNA in a way that completely bypasses eggs and sperm. In a sense, her nurturing behavior tells her pups something about the world they will grow up in. Mom's behavior actually programs the pups' DNA in a way that will make them more likely to succeed.

The epigenetic code gives the genome a level of flexibility that extends beyond the relatively fixed DNA code. The epigenetic code allows certain types of information to be passed to offspring without having to go through the slow processes of random mutation and natural selection. At the same time, the epigenetic code is sensitive to changing environmental conditions such as availability of food or threat from predators.

High-nurturing mothers raise high-nurturing offspring, and low-nurturing mothers raise low-nurturing offspring. This may look like a genetic pattern, but it's not. Whether a pup grows up to be anxious or relaxed depends on the mother that raises it - not the mother that gives birth to it.

EPIGENETIC PATTERNS ARE REVERSIBLE

Gene expression patterns that are set up early in life are not necessarily stuck that way forever. You can take a low-nurtured rat, inject its brain with a drug that removes methyl groups, and make it act just like a high-nurtured rat. The GR gene gets turned on, cells make more GR protein, and the rat takes on a more relaxed personality. It works in the other direction too. You can take a relaxed, high-nurtured rat, inject its brain with methionine (a source of methyl) and make it more anxious. Of course drugs affect many genes, so they're not an exact substitute for maternal care. But it turns out that you can also turn an anxious rat into a more relaxed rat by spicing up its living quarters. So take heart — your epigenetic destiny is not written in permanent ink.

<u>Dr. Moshe Szyf</u>, Professor of Pharmacology and Therapeutics at McGill University, talks about high and low nurturing rats.

In a 2004 article, researchers at McGill University gave us some of the first clues about how social interactions help to shape the epigenome. The study showed that there are epigenetic differences between high- and low-nurtured rats.

When we're confronted with danger, the body turns on stress circuitry in the brain. Stress circuitry activates the adrenaline-driven Fight or Flight response and causes the hormone cortisol to be released into the bloodstream. Cortisol is important for freeing stored energy, which helps with both fighting and fleeing. But too much cortisol can be a bad thing. High levels can lead to heart disease, depression, and increased susceptibility to infection. Cortisol also travels to an area of the brain called the hippocampus, where it binds to GRs. When enough cortisol is bound, the hippocampus sends out signals that turn off the stress circuit, shutting down both the Fight or Flight response and cortisol production.

Rats (and people) with higher levels of GR are better at detecting cortisol, and they recover from stress more quickly. Stress signals travel from the hypothalamus to the pituitary gland and then to the adrenal glands. The adrenal glands release the hormone cortisol (and adrenaline, not shown).

When cells in the hippocampus detect cortisol, which binds to the GR receptor, they send a signal to the hypothalamus that shuts down the stress circuit.

Our genes are turned on or off, even before we were born. [12] When we are born, the deoxyribonucleic acid/DNA in our bodies contains the blueprints for who we are and instructions for who we will become.

Many people have mistakenly believed that the DNA with which we are born is the sole determinant for who we are and will become, but scientists have understood for decades that this *genetic determinism* is a flawed theory.

The field of *epigenetics* refers to the science that studies how the development, functioning and evolution of biological systems are influenced by forces operating outside the DNA sequence, including intracellular, environmental and energetic influences.

Since the 1950s scientists have accepted that epigenetic influence is critical in our development. *Epi* – Greek for "besides" – combines with the word, *genetics*, to essentially mean "something more than genetics." That "something more" is widely held today to refer to our environment – thus meaning that our genetic code and the environment in which we develop determine who and what we are.

Researchers have shown through studies that epigenetics entails even more than DNA and the places where we live, the climate around us and all the twists, turns and hard knocks of our lives.

Stem cell biologist and bestselling author Bruce Lipton, Ph.D. says the distinction between genetic determinism and epigenetics is important.[12]

Our emotions are a reflection of our gut function, so whatever's going on in your gastrointestinal (GI)tract may be a true reflection of what's going on in your head and vice versa. [21]

In simple terms most people can relate to, what this means is that when we are having a bad day, going through a rough period such as dealing with the sickness of a loved one or coping with financial troubles, we can actually influence our bodies – all the way down to the cellular level – by intentionally thinking positive thoughts and focusing on positive emotions.

DISCUSSION

The power of intentional thoughts and emotions goes beyond theory at the HeartMath Institute. In a study, researchers have tested this idea and proven its veracity. HeartMath researchers have gone so far as to show that physical aspects of DNA strands could be influenced by human intention. The article, Modulation of DNA Conformation by Heart-Focused Intention – McCraty, Atkinson, Tomasino, 2003 – describes experiments that achieved such results.[22]

For example, an individual holding three DNA samples was directed to generate *heart coherence* – a beneficial state of mental, emotional and physical balance and harmony – with the aid of a HeartMath technique that utilizes heart breathing and intentional positive emotions. The individual succeeded, as instructed, to intentionally and simultaneously unwind two of the DNA samples to different extents and leave the third unchanged.

"The results provide experimental evidence to support the hypothesis that aspects of the DNA molecule can be altered through intentionality," the article states. "The data indicate that when individuals are in a heart-focused, loving state and in a more coherent mode of physiological functioning, they have a greater ability to alter the conformation of DNA.

"Individuals capable of generating high ratios of heart coherence were able to alter DNA conformation according to their intention. ... Control group participants showed low ratios of heart coherence and were unable to intentionally alter the conformation of DNA."

The influence or control individuals can have on their DNA – who and what they are and will become – is further illuminated in HeartMath founder Doc Childre's theory of *heart intelligence*. Childre postulates that "an energetic connection or coupling of information" occurs between the DNA in cells and higher dimensional structures – the higher self or spirit.

Childre further postulates, "The heart serves as a key access point through which information originating in the higher dimensional structures is coupled into the physical human system (including DNA), and that states of heart coherence generated through experiencing heartfelt positive emotions increase this coupling."

The heart, which generates a much stronger electromagnetic field than the brain's, provides the energetic field that binds together the higher dimensional structures and the body's many systems as well as its DNA.

Childre's theory of heart intelligence proposes that "individuals who are able to maintain states of heart coherence have increased coupling to the higher dimensional structures and would thus be more able to produce changes in the DNA." [23] Now fascinating new research is showing stem-cell programming may not be restricted to chemical signaling, that it's possible it also can be achieved by extremely low frequency signals in electromagnetic fields similar to the fields produced by the human heart and Earth. We can learn more about this by HeartMath Institute Director of Research Dr. Rollin McCraty. [24]

CONCLUSIONS

Development of a process akin to a "time machine" that is capable of reprogramming adult cells into other types of cells.

A finding that cells can produce acoustic/sound vibrations and the ability of cells to express vibrational signatures, which change depending on the task the cells are performing and which reflect the cells' health.

The possibility that sounds emitted by organs contain information that regulates cellular functions at molecular, sub molecular, and quantum levels.

Nobel Laureate Luc Montagnier's revolutionary experiment in which he purports that a virtually identical copy of a fragment of DNA in one test tube was "teleported" via electromagnetic signals to a second test tube containing pure water.

New research being conducted to determine if the results of Montagnier's DNA teleport experiments can be recreated at the whole cell level in stem-cell cultures.

New finding is clear example in humans of the theory of epigenetic inheritance: the idea that environmental factors can affect the genes of your children

Epigenetics patterns are reversible. Genes are turned on and off by modifications to the tail of histones, such as acetylation.

Let´s hope that our good genes are turned on and not will be turned off.

[¹] Fredriksson, I. The Journey to Life or Death. USA/Singapore: Strategic Book Publishing and Rights Co., LLC; 2015. p.119.

[²] Wilhelmsson, P. Lev ung längre (Live Young Longer) pp.33. Västerås: ICA bokförlag; 2000. Se also L. Hayflick, How and Why We Age, Ballantie 1994, p.xxix.

[³] Nobelförsamlingen, The Nobel Assembly at Karolinska Institutet. Maintenance of chromosomes by telomeres and the enzyme telomeras. [cited 2016 Nov. 24] Available from: http://www.nobelprize.org/nobel_prizes/medicine/laureates/2009/advanced-medicineprize2009.pdf

[⁴] Azvolinsky A. Exercise Boosts Telomere Transcription, Endurance exercise and metabolism are linked to transcriptional activation of human telomeres, researchers propose. [cited 2016 Nov. 24] Available from: http://www.thescientist.com/?articles.view/articleNo/46657/title/Exercise-Boosts-Telomere-Transcription/

[⁵] What happens when DNA becomes damaged? [cited 2016 Nov. 26] Available from: https://www.nia.nih.gov/health/publication/genetics-aging-our-genes/what-happens-when-dna-becomes-damaged

[⁶] Panchision D M. Reparing the Nervous System with Stem Cells. [cited 2016 Nov. 24] Available from: https://stemcells.nih.gov/info/Regenerative_Medicine/2006Chapter3.htm

[⁷] Leonardo Mariño-Ramírez, Maricel G Kann, Benjamin A Shoemaker, and David Landsman Histone structure and nucleosome stability PMCID: PMC1831843 NIHMSID: NIHMS17286 Expert Rev Proteomics. 2005 Oct; 2(5): 719–729. doi: 10.1586/14789450.2.5.719

[8] Non–Coding RNA. [cited 2016 Nov. 26] Available from: http://www.whatisepigenetics.com/non-coding-rna/

[9] What is Epigenetics. [cited 2016 Nov. 26] Available from: http://www.whatisepigenetics.com/dna-methylation/

[10] Haglund O. Den viktiga epigenetiken och hjärtsjukdom, (The important epigenetics and heartdisease). Medicinsk Access 2013; nr 2: 17-23.

[11] Schéle E, Grahnemo L Anesten F, Hallén, A. Bäckhed F, Jansson J-O. The Gut Microbiota Reduces Leptin Sensitivity and the Expression of the Obesity-Suppressing Neuropeptides Proglucagon (Gcg) and Brain-Derived Neurotrophic Factor (Bdnf) in the Central Nervous System. [cited 2016 Nov. 24] Available from: http://press.endocrine.org/doi/full/10.1210/en.2012-2151

[12] Fredriksson, I. The Journey to Life or Death. USA/Singapore: Strategic Book Publishing and Rights Co., LLC; 2015. p 126.

[13] Epigenetics of Obesity. Differential expression of a chromatin-interacting protein is linked to weight variation in mice and humans, researchers show. http://www.the-scientist.com/?articles.view/articleNo/45199/title/Epigenetics-of-Obesity/By Catherine | January 29, 2016

[14] Offord C. Epigenetic switch for obesity, Obesity can sometimes be shut down. January 28, 2016 [cited 2016 Nov. 24] Available from: http://www.thescientist.com/?articles.view/articleNo/45199/title/Epigenetics-of Obesity/

[15] Hansen J. How obesity makes memory go bad. [cited 2016 Nov. 24] Available from: https://www.uab.edu/medicine/news/latest/item/943-how-obesity-makes-memory-go-bad

[16] The Dutch famine birth cohort study. [cited 2016 Nov. 24] Available from: http://www.hongerwinter.nl/item.php?id=32&language=EN

1[17]] Birney E. Study of Holocaust survivors finds trauma passed on to children's genes. [cited 2016 Nov. 24] Available from: https://www.theguardian.com/science/2015/aug/21/study-of-holocaust-survivors-finds-trauma-passed-on-to-childrens-genes

[18] HeartMate Institute. You Can Change Your DNA. [cited 2016 Nov. 24] Available from: https://www.heartmath.org/articles-of-the-heart/personal-development/you-can-change-your-dna/

[19] Mum's in control – even before you're born. [cited 2016 Nov. 24] Available from: http://www.babraham.ac.uk/news/2016/01/egg_epigenetic_blueprint

[20] Lick your rats. [cited 2016 Nov. 24] Available from: http://learn.genetics.utah.edu/content/epigenetics/rats/

[21] What's the Connection Between Your Brain and Your Gut?. [cited 2016 Nov. 27] Available from: http://blog.myneurogym.com/connection-between-your-brain-and-gut

[22] McCraty R, Atkinson M, Tomasino Modulation of DNA Conformation By Heart-Focused Intention. [cited 2016 Nov. 27] Available from: http://www.aipro.info/drive/File/224.pdf

[23] Henry D. Why your genes don't have the final word on your health. [cited 2016 Nov. 24] Available from: http://biosilblog.com/tag/epigenetics/

[24] Hershey S. The Coherent Heart: A Discussion with Dr. Rollin McCraty [cited 2016 Nov. 24] Available from: http://realitysandwich.com/130470/the_coherent_heart/

ABOUT THE CONTRIBUTORS

RICHARD L. AMOROSO IS Professor and a theoretical physicist is also a Mormon High Priest and claims to be the first true Noeticist. He is Director of the Noetic Advanced Studies Institute, and has been Chairman of the international symposium series in honor of French mathematical physicist Jean-Pierre Vigier for the past twenty years. He is author of over thirty volumes and several hundred papers/chapters, some of which are translated into five languages. He is honorary member of the Romanian Academy of Science. His most recent monograph is on Universal Quantum Computing. Amoroso is a prolific writer in numerous fields of physics, medicine, philosophy of mind, epigenetics and theology. His major contribution is a comprehensive solution to the Mind-Body problem, not yet popular because it is Cartesian, in opposition to current thinking. However, he is involved in preparing an empirical test, which will test for higher dimensions beyond the Standard Model; which if successful will also put an end to the need for massive accelerators like CERN.

Gerard J.F. Blommestijn received his masters degree in experimental physics in 1971 and his PhD in 1980, both from the University of Amsterdam. From 1971 till 1980 he was staff member in the Institute for Nuclear Physics. During that time he also studied the interpretation of quantum mechanics as well as oriental scriptures. From 1980 till 1987 he developed software. From 1987, he was first head of the Biophysics Department and later software engineer at the Netherlands Cancer Institute in Amsterdam. He then also developed a hypothesis on the relation between the quantum mechanical reduction process and consciousness.

Carl Johan Calleman holds a PhD in Physical Biology from the University of Stockholm and has among other things served as a Senior Researcher of Environmental Health at the University of Washington an a cancer expert for the WHO. He is currently Professor of Chemistry at the Santa Fe Community College.

Ingrid Fredriksson is a Swedish author with an M.A. degree in public health education. She has written several books, including Flow Forever, The Third Book, The Power of Thought, Free from Dangerous Stress, H2O Just Ordinary Water, There is No Death, The Journey to Life or Death and a e-book for children Rose – The Lucky Cow of Small Meadow and she is editor for Aspects of Consciousness Essays on Physics, Death and the Mind, and for The Mysteries of Consciousness Essays on Spacetime, Evolution and Well-Being.

Attila Grandpierre is a professional physicist and astrophysicist as well as an historian and an accomplished musician, having been the leader and lead singer of several bands starting in the 70s and until the present day. Attila Grandpierre is a well-known and active figure in Hungary as a researcher of the ancient Magyar tradition, history, myth and spirituality.He studied theoretical biology focused on Ervin Bauer's works. In 2009, his subject field of interest concerned the relation between astronomy and civilization. As physicist, he had a strong interest in the problem of bringing the sciences and metaphysics together. He paid special attention to interdisciplinal science and complexity of living systems in 2008. As a physicist, Grandpierre published 10 books, 3 book chapters, more than 70 science and over 300 popular science papers. In astrophysics, he wrote an article on the variable nature of the Sun's core, which was mentioned in New Scientist in 2007.

Alexander J. Graur was born 1952 in Romania. Italian citizen. Dr. Alexander J. Graur, Ph.D. Member The New York Academy of Sciences Project Associate, School of Design The Hong Kong Polytechnic University. Between 2001-2005 lived and worked in the USA (New Jersey). Actually living in Italy, Turin. Degree in Composition/Musicology; undergraduate, School of Medicine/Psychiatry, University of Bucharest, Romania. Board Certified Music Therapist by CBMT (USA, 2003/renewed 2008). Member of American Music Therapists Association since 2003. Founder and president, Medicamus Italiana. Torino, Italy, (former Medicamus Center LLC, Edgewater, NJ, USA). Founder (2008), International Encounters in Clinical Music Therapy, Piedmont, Italy. Works in composition, musicology, music therapy published in Italy and USA; recordings as soloist and composer published in Romania, Italy, France, Germany. National prizes in composition and interpretation in Italy, Romania, USA, UK.

Since 1978 developed the Music Integrative Neurotherapy, TM, an Applied Neuroscience Therapy, combining the rules of the Music Composition with the Medicine, Molecular Biology and Physics. The method is used in Psychiatric and Neurological environments as a Complementary Medicine therapy. This method and its uses was presented at international conferences in the USA, Korea, China, Italy.

For more details, please visit: www.medicamus.com

Eve Isham. Ph. D. (USA) Saving Free Will from Science. Eve Isham is an assistant professor at University of Arizona. She approaches the scientific study of consciousness from the perspective of a cognitive psychologist. Her research interests include time perception, intentionality, and decision making.

Olle Johansson associate professor at the Karolinska Institute, Department of Neuroscience, and head of The Experimental Dermatology Unit, has a long background in the neurosciences and has coauthored – together with his supervisor professor Tomas Hökfelt and many others, including Nobel Laureates – up to the presentation of his doctoral thesis 143 original papers, reviews, book chapters and conference abstracts, a publication record hard to beat! His doctoral thesis at the Karolinska Institute was entitled "Peptide Neurons in the Central and Peripheral Nervous System. Light and Electron Microscopic Studies." Olle Johansson is a world-leading authority in the field of EMF radiation and health effects. Among many achievements, he coined the term "screen dermatitis" which later on was developed into the functional impairment electrohypersensitivity which recognition mainly is due to his work.

Rupert Sheldrake, Ph. D. (England) is a biologist and author, is best known for his hypothesis of morphic fields and morphic resonance, which leads to a vision of a living, developing universe with its own inherent memory. He worked in developmental biology at Cambridge University, where he was a Fellow of Clare College. He was then Principal Plant Physiologist at the International Crops Research Institute for the Semi-Arid Tropics (ICRISAT), in Hyderabad, India. From 2005 to 2010, he was Director of the Perrott-Warrick project, funded from Trinity College, Cambridge

Klaus Stüben, Ph. D. (Germany), is a pharmacist with major interests in the functioning of the brain under different situations and a scientist with the focus on consciousness itself. Since his seventeenth birthday, he has been interested in aspects of consciousness and in the search for freedom and happiness. He did a lot of self inquires, experiments on his own and with groups of people. He worked as a medical practitioner and body therapist and supports people in his trainings and courses individually or in groups. He is supporting them with great joy in their direct experience of consciousness with all aspects. He defined innerchi – direct experience as a doorway to consciousness. As he is trained as physiotherapist, he is using a lot of different skills such as NLP, trance work, hypnosis therapy as well as a wealth of experience in dealing with physical, mental and emotional energies. Based on his experiences he wrote several articles and the book: Klaus Stüben, Du bist frei und kannst das Jetzt direkt erleben Windpferd Verlag 2010. (You are free) www.innerchi.de

G Anita K Westlund was born 1944. She is engineer, teacher and philosopher and is very grateful for her exciting life. Early manifested with a patent as her company sold to a concurrent.

For her it's obvious the big humanity challenge is about accepting a holographic consciousness in us, decoding the quantum flow into a new level of life and peace and technology to came through. And there is no short ways to heaven but through the population involved.

She has supplied for about two hundred points at the university, except for her professions and including master studies in didactics, as alternatives of about the same. 2002-03 she was greatly engaged against the EMF-mast system to be builded up in our specially protected area. Most threatening young and inner sensitive ones. Rendering her the Värmlandian environmental prize of the year.

Not to forget the artistical outlets and her playing piano through the years. She has also written two books in Swedish, the first is Kvanthopp – se människan! (Quantum hope- see the human) 2011, and 2013 Naturens gyllene snitt och Cirkelns Kvadratur, en andlig utmaning (Holy Scales in the geografics, a spiritual challenge). Her books are the fact behind her article in this book.

Printed in the United States
By Bookmasters

Printed in the United States
By Bookmasters